ASTRONOMICAL PHENOMENA

FOR THE YEAR

2007

Prepared Jointly by
The Nautical Almanac Office
United States Naval Observatory

and

Her Majesty's Nautical Almanac Office
United Kingdom

WASHINGTON
U.S. Government Printing Office

2004

UNITED STATES

Printed in the United States of America
by the U. S. Government Printing Office
by permission

For sale by the
U.S. Government Printing Office
Superintendent of Documents
P. O. Box 371954
Pittsburgh, PA 15250-7954
phone: 1-202-512-1800
order online at http://bookstore.gpo.gov/

Available from
HM Nautical Almanac Office
Rutherford Appleton Laboratory
Chilton
Didcot
Oxon OX11 0AX
hmnao@nao.rl.ac.uk

Further information:
http://aa.usno.navy.mil/AA/
http://www.nao.rl.ac.uk/
http://www.earthandsky.co.uk/

ASTRONOMICAL PHENOMENA

FOR THE YEAR 2007

CONTENTS

The astronomical data in this booklet are expressed in the scale of universal time (UT); this is also known as Greenwich mean time (GMT) and is the standard time of the Greenwich meridian (0° of longitude). A time in UT may be converted to local mean time by the addition of east longitude (or subtraction of west longitude), where the longitude of the place is expressed in time-measure at the rate of 1 hour for every 15°. The differences between standard times and UT are indicated in the chart on page 22; local clock times may, however, differ from these standard times, especially in summer when clocks are often advanced by 1 hour.

PRINCIPAL PHENOMENA OF SUN AND MOON, 2007

THE SUN

		d h			d h m		d h m
Perigee	... Jan.	3 20	Equinoxes	... Mar.	21 00 07	... ... Sept.	23 09 51
Apogee	... July	7 00	Solstices	... June	21 18 06	... ... Dec.	22 06 08

PHASES OF THE MOON

Lunation	New Moon		First Quarter		Full Moon		Last Quarter	
		d h m		d h m		d h m		d h m
1039					Jan.	3 13 57	Jan.	11 12 45
1040	Jan.	19 04 01	Jan.	25 23 01	Feb.	2 05 45	Feb.	10 09 51
1041	Feb.	17 16 14	Feb.	24 07 56	Mar.	3 23 17	Mar.	12 03 54
1042	Mar.	19 02 43	Mar.	25 18 16	Apr.	2 17 15	Apr.	10 18 04
1043	Apr.	17 11 36	Apr.	24 06 36	May	2 10 09	May	10 04 27
1044	May	16 19 27	May	23 21 03	June	1 01 04	June	8 11 43
1045	June	15 03 13	June	22 13 15	June	30 13 49	July	7 16 54
1046	July	14 12 04	July	22 06 29	July	30 00 48	Aug.	5 21 20
1047	Aug.	12 23 03	Aug.	20 23 54	Aug.	28 10 35	Sept.	4 02 32
1048	Sept.	11 12 44	Sept.	19 16 48	Sept.	26 19 45	Oct.	3 10 06
1049	Oct.	11 05 01	Oct.	19 08 33	Oct.	26 04 52	Nov.	1 21 18
1050	Nov.	9 23 03	Nov.	17 22 33	Nov.	24 14 30	Dec.	1 12 44
1051	Dec.	9 17 40	Dec.	17 10 18	Dec.	24 01 16	Dec.	31 07 51

ECLIPSE

A total eclipse of the Moon	Mar. 3-4	The Arctic, Asia except the eastern part, Europe including the British Isles, Africa, South America and the eastern parts of central and North America.
A partial eclipse of the Sun	Mar. 19	Most of Alaska, eastern and central Asia except the central parts of Japan and the western part of Russia.
A total eclipse of the Moon	Aug. 28	Americas except the eastern part of South America and the north-eastern parts of North America, the Pacific Ocean, eastern parts of Asia, Australasia and Antarctica.
A partial eclipse of the Sun	Sept. 11	Parts of Antarctica, South America except the northern part and the south-western Atlantic Ocean.

For further details see pages 65–73

MOON AT PERIGEE

	d h		d h		d h
Jan.	22 13	June	12 17	Oct.	26 12
Feb.	19 10	July	9 22	Nov.	24 00
Mar.	19 19	Aug.	4 00	Dec.	22 10
Apr.	17 06	Aug.	31 00		
May	15 15	Sept.	28 02		

MOON AT APOGEE

	d h		d h		d h
Jan.	10 16	May	27 22	Oct.	13 10
Feb.	7 13	June	24 14	Nov.	9 13
Mar.	7 04	July	22 09	Dec.	6 17
Apr.	3 09	Aug.	19 03		
Apr.	30 11	Sept.	15 21		

OCCULTATIONS OF PLANETS AND BRIGHT STARS BY THE MOON

Date (d h)	Body	Areas of Visibility
Jan. 6 18	Saturn	N.E. Russia, Arctic regions, N. Scandinavia, N.W. Canada, Alaska
Jan. 7 05	*Regulus*	E. Europe, E. Scandinavia, W. Russia
Jan. 11 20	*Spica*	Most of Antarctica, S.E. Indian Ocean
Jan. 15 13	*Antares*	S. tip of Africa, part of Antarctica, S. tip of S. America
Jan. 20 17	Venus	S.W. Africa, most of Antarctica, S. tip of S. America
Jan. 22 06	Uranus	Japan, Philippines, Indonesia, E. Indian Ocean, S. tip of India
Feb. 02 23	Saturn	Central Asia, E. Scandinavia, Arctic regions
Feb. 03 14	*Regulus*	N.W. North America, N. Greenland
Feb. 08 04	*Spica*	Drake Passage, S. of S. America
Feb. 11 22	*Antares*	Southern oceans, Antarctica
Mar. 02 02	Saturn	W. Russia, Europe except W. British Isles and S.W. Europe
Mar. 02 21	*Regulus*	E. Central Asia, Arctic regions
Mar. 11 06	*Antares*	Antarctica, S. part of S. America
Mar. 17 03	Mercury	Southern Ocean south of New Zealand
Mar. 29 04	Saturn	N. British Isles, N. Scandinavia, N. Atlantic Ocean, E. Greenland
Mar. 30 03	*Regulus*	W. Europe inc. British Isles, Scandinavia, Arctic regions
Apr. 07 13	*Antares*	S. part of S. America, W. Antarctica, W. Oceania
Apr. 14 02	Mars	S. and E. Asia, India, E. tip of Africa
Apr. 14 20	Uranus	E. Siberia, Japan, Alaska, N.W. Canada
Apr. 25 10	Saturn	N. Greenland, N.W. Canada, Alaska, E. tip of Siberia
Apr. 26 09	*Regulus*	N.W. North America, Arctic regions
May 04 18	*Antares*	New Zealand, Tasmania, par of Antarctica, S.E. Africa
May 12 07	Uranus	N. Atlantic Ocean, British Isles except S.E. part, E. Greenland
May 22 19	Saturn	Europe inc. British Isles, N.E. Africa, N.W. Asia, Arctic regions, N.W. Canada
May 23 16	*Regulus*	Asia except E. part, N.E. Europe including British Isles, Greenland, N.E. tip of Canada
June 01 01	*Antares*	S. half of S. America, part of Antarctica, S.W. Indian Ocean
June 18 15	Venus	W. Asia, Europe including British Isles except S. Iberia, Greenland, N. Canada
June 19 08	Saturn	Japan, Central Asia, E. part of Europe
June 20 00	*Regulus*	E. Siberia, N. America except N.E., Caribbean, N.W. South America
June 28 08	*Antares*	W. Oceania, W. tip of Antarctica, S. part of S. America
July 03 19	Neptune	Bellingshausen Sea (Antarctica)
July 16 23	Saturn	Hawaiian Islands, W. parts of central S. America
July 17 09	*Regulus*	Europe including British Isles, S. and W. Asia, Indonesia, S. Philippines, N.W. Australia
July 25 16	*Antares*	S. tip of Africa, most of Antarctica, S. parts of Australia and New Zealand
July 31 01	Neptune	Part of Antarctica, Kerguelen Is
Aug. 22 01	*Antares*	Antarctica, southern oceans, New Zealand
Sept. 10 01	*Regulus*	Polynesia, Japan, Central Asia
Sept. 10 04	Saturn	S. Indian Ocean, W. tip of Australia, part of Antarctica
Sept. 18 08	*Antares*	Antarctica, southern oceans, S. Madagascar
Oct. 07 07	*Regulus*	Europe inc. S. British Isles, N. and E. Africa, Middle East
Oct. 07 16	Saturn	Southern Ocean S. of Polynesia
Oct. 15 15	*Antares*	Larger part of Antarctica, S. half of S. America
Oct. 21 03	Neptune	Part of Antarctica, S. Georgia
Nov. 03 13	*Regulus*	S. North America, Caribbean, north S. America
Nov. 11 21	*Antares*	S. part of S. America, S. Pacific Ocean, most of New Zealand, Polynesia
Nov. 17 11	Neptune	Antarctica, S. Australia, New Zealand

continued on page 7 . . .

GEOCENTRIC PHENOMENA

MERCURY

	d h	d h	d h	d h
Superior conjunction . . .	Jan. 7 06	May 3 04	Aug. 15 20	Dec. 17 15
Greatest elongation East	Feb. 7 17 (18°)	June 2 10 (23°)	Sept. 29 16 (26°)	—
Stationary	Feb. 13 14	June 15 16	Oct. 12 07	—
Inferior conjunction . . .	Feb. 23 05	June 28 19	Oct. 24 00	—
Stationary	Mar. 7 10	July 10 02	Nov. 1 13	—
Greatest elongation West	Mar. 22 02 (28°)	July 20 15 (20°)	Nov. 8 21 (19°)	—

VENUS

	d h		d h
Greatest elongation East	June 9 03 (45°)	Stationary	Sept. 7 14
Greatest illuminated extent	July 12 14	Greatest illuminated extent	Sept.23 23
Stationary	July 25 13	Greatest elongation West	Oct. 28 15 (46°)
Inferior conjunction . . .	Aug. 18 04		

EARTH

	d h		d h m		d h m
Perihelion . . .	Jan. 3 20	Equinoxes . . .	Mar. 21 00 07 . . .	. . .	Sept. 23 09 51
Aphelion . . .	July 7 00	Solstices . . .	June 21 18 06 . . .	. . .	Dec. 22 06 08

SUPERIOR PLANETS

	Stationary	Opposition	Stationary	Conjunction
	d h	d h	d h	d h
Mars	Nov. 15 16	Dec. 24 20	—	—
Jupiter	Apr. 6 02	June 5 23	Aug. 7 06	Dec. 23 06
Saturn	Dec. 20 12 |	Feb. 10 19	Apr. 20 01	Aug. 21 23
Uranus	June 23 23	Sept. 9 19	Nov. 24 18 |	Mar. 5 16
Neptune	May 25 06	Aug. 13 18	Oct. 31 20 |	Feb. 8 16
Pluto	Apr. 1 00	June 19 07	Sept. 7 22	Dec. 21 00

The vertical bars indicate where the dates for the planet are not in chronological order.

HELIOCENTRIC PHENOMENA

	Perihelion	Aphelion	Greatest Lat. South	Ascending Node	Greatest Lat. North	Descending Node
Mercury	Feb. 9	Mar. 25	Jan. 17	Feb. 5	Feb. 20	Mar. 15
	May 8	June 21	Apr. 15	May 4	May 19	June 11
	Aug. 4	Sept. 17	July 12	July 31	Aug. 15	Sept. 7
	Oct. 31	Dec. 14	Oct. 8	Oct. 27	Nov. 11	Dec. 4
Venus	Apr. 19	Aug. 9	Jan. 19	Mar. 16	May 11	July 6
	Nov. 30	—	Aug. 31	Oct. 27	Dec. 21	—
Mars	June 4	—	May 9	Oct. 3	—	—

Uranus: Greatest Lat. South, Jan. 3

Jupiter, Saturn, Neptune, Pluto: None in 2007

VISIBILITY OF PLANETS

MERCURY can only be seen low in the east before sunrise, or low in the west after sunset (about the time of beginning or end of civil twilight). It is visible in the mornings between the following approximate dates: March 1 to April 25, July 7 to August 8, and October 30 to December 1. The planet is brighter at the end of each period, (the best conditions in northern latitudes occur in the first half of November and in southern latitudes from mid-March to early April). It is visible in the evenings between the following approximate dates: January 21 to February 17, May 11 to June 20 and August 25 to October 18. The planet is brighter at the beginning of each period, (the best conditions in northern latitudes occur in the first half of February and mid-May to early June and in southern latitudes from mid-September to mid-October).

VENUS is a brilliant object in the evening sky until mid-August when it becomes too close to the Sun for observation. From the start of the fourth week in August it reappears in the morning sky where it stays until the end of the year. Venus is in conjunction with Saturn on July 2, August 9 and October 15.

MARS can be seen in the morning sky from the beginning of the year as it passes through Ophiuchus, Sagittarius, Capricornus, Aquarius, Pisces, briefly into Cetus, returning to Pisces then into Aries and Taurus. Its westward elongation increases (passing 5° N of *Aldebaran* on August 24) and moves into Gemini in late September. It is at opposition on December 24 when it is visible throughout the night.

JUPITER rises well before sunrise at the beginning of the year in Ophiuchus (passing 5° N of *Antares* on January 5) and by early March can be seen for more than half the night. Its westward elongation gradually increases until on June 5 it is at opposition when it is visible throughout the night. Its eastward elongation then decreases and from early September can only be seen in the evening sky. It passes into Sagittarius in early December but during the second week of December it becomes too close to the Sun for observation.

SATURN rises after sunset at the beginning of the year in Leo and remains in this constellation throughout the year. It is at opposition on February 10 when it can be seen throughout the night. From mid-May until early August it is visible only in the evening sky, and then becomes too close to the Sun for observation. It can be seen only in the morning sky from the start of the second week of September until early December. For the remainder of the year it is visible for more than half the night. Saturn is in conjunction with Venus on July 2, August 9 and October 15.

URANUS is visible from the beginning of the year until mid-February in the evening sky in Aquarius and remains in this constellation throughout the year. From mid-February it becomes too close to the Sun for observation and reappears in late March in the morning sky. It is at opposition on September 9. Its eastward elongation gradually decreases and from early December it can be seen only in the evening sky.

NEPTUNE is visible for the first half of January in the evening sky in Capricornus and remains in this constellation throughout the year. It then becomes too close to the Sun for observation until early March when it reappears in the morning sky. It is at opposition on August 13 and from mid-November can be seen only in the evening sky.

DO NOT CONFUSE (1) Venus with Saturn in late June to mid-July and again with Saturn in mid-October on both occasions Venus is the brighter object.

VISIBILITY OF PLANETS IN MORNING AND EVENING TWILIGHT

	Morning	Evening
Venus	August 22 – December 31	January 1 – August 13
Mars	January 1 – December 24	December 24 – December 31
Jupiter	January 1 – June 5	June 5 – December 10
Saturn	January 1 – February 10 September 9 – December 31	February 10 – August 4

... continued from page 5

OCCULTATIONS OF PLANETS AND BRIGHT STARS BY THE MOON

Date (d h)	Body	Areas of Visibility
Nov. 30 20	*Regulus*	E. India, S.E. Asia, N.E. Australia, New Zealand, Micronesia, Melanesia
Dec. 12 21	Vesta	S. America except N.W. part, New Zealand
Dec. 14 18	Neptune	Part of Antarctica, S. tip S. America, S. Africa
Dec. 16 01	Pallas	N.W. Canada, Alaska, E. tip of Siberia, Hawaiian Islands
Dec. 24 03	Mars	N.W. Canada, Alaska, Arctic regions, N. Russia, E. Europe, N.E. British Isles
Dec. 28 05	*Regulus*	S. America except S. part, S. Atlantic Ocean

VISIBILITY OF PLANETS

The planet diagram on page 9 shows, in graphical form for any date during the year, the local mean times of meridian passage of the Sun, of the five planets, Mercury, Venus, Mars, Jupiter and Saturn, and of every 2^h of right ascension. Intermediate lines, corresponding to particular stars, may be drawn in by the user if desired. The diagram is intended to provide a general picture of the availability of planets and stars for observation during the year.

On each side of the line marking the time of meridian passage of the Sun, a band 45^m wide is shaded to indicate that planets and most stars crossing the meridian within 45^m of the Sun are generally too close to the Sun for observation.

For any date the diagram provides immediately the local mean time of meridian passage of the Sun, planets and stars, and thus the following information:
a) whether a planet or star is too close to the Sun for observation;
b) visibility of a planet or star in the morning or evening;
c) location of a planet or star during twilight;
d) proximity of planets to stars or other planets.

When the meridian passage of a body occurs at midnight, it is close to opposition to the Sun and is visible all night, and may be observed in both morning and evening twilights. As the time of meridian passage decreases, the body ceases to be observable in the morning, but its altitude above the eastern horizon during evening twilight gradually increases until it is on the meridian at evening twilight. From then onwards the body is observable above the western horizon, its altitude at evening twilight gradually decreasing, until it becomes too close to the Sun for observation. When it again becomes visible, it is seen in the morning twilight, low in the east. Its altitude at morning twilight gradually increases until meridian passage occurs at the time of morning twilight, then as the time of meridian passage decreases to 0^h, the body is observable in the west in the morning twilight with a gradually decreasing altitude, until it once again reaches opposition.

Notes on the visibility of the principal planets, except Pluto, are given on page 7. Further information on the visibility of planets may be obtained from the diagram below which shows, in graphical form for any date during the year, the declinations of the bodies plotted on the planet diagram on page 9.

DECLINATION OF SUN AND PLANETS, 2007

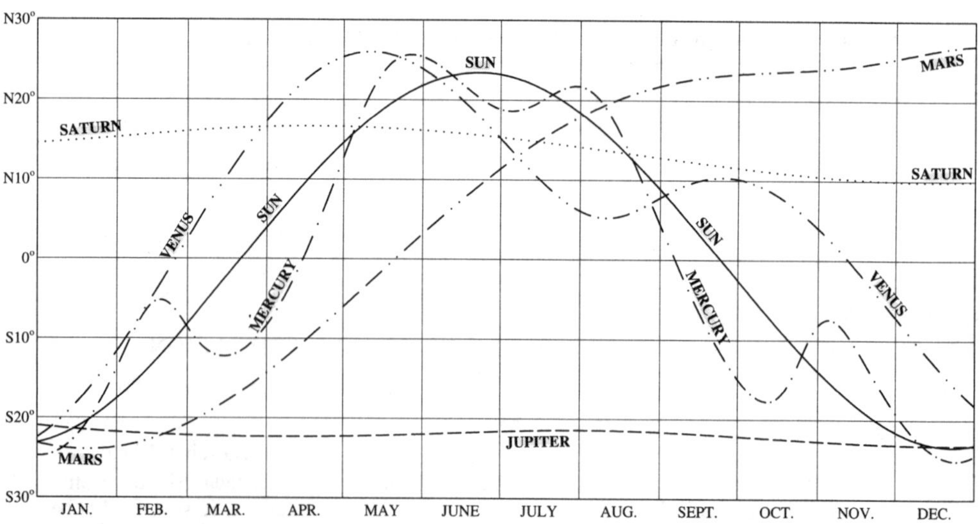

LOCAL MEAN TIME OF MERIDIAN PASSAGE

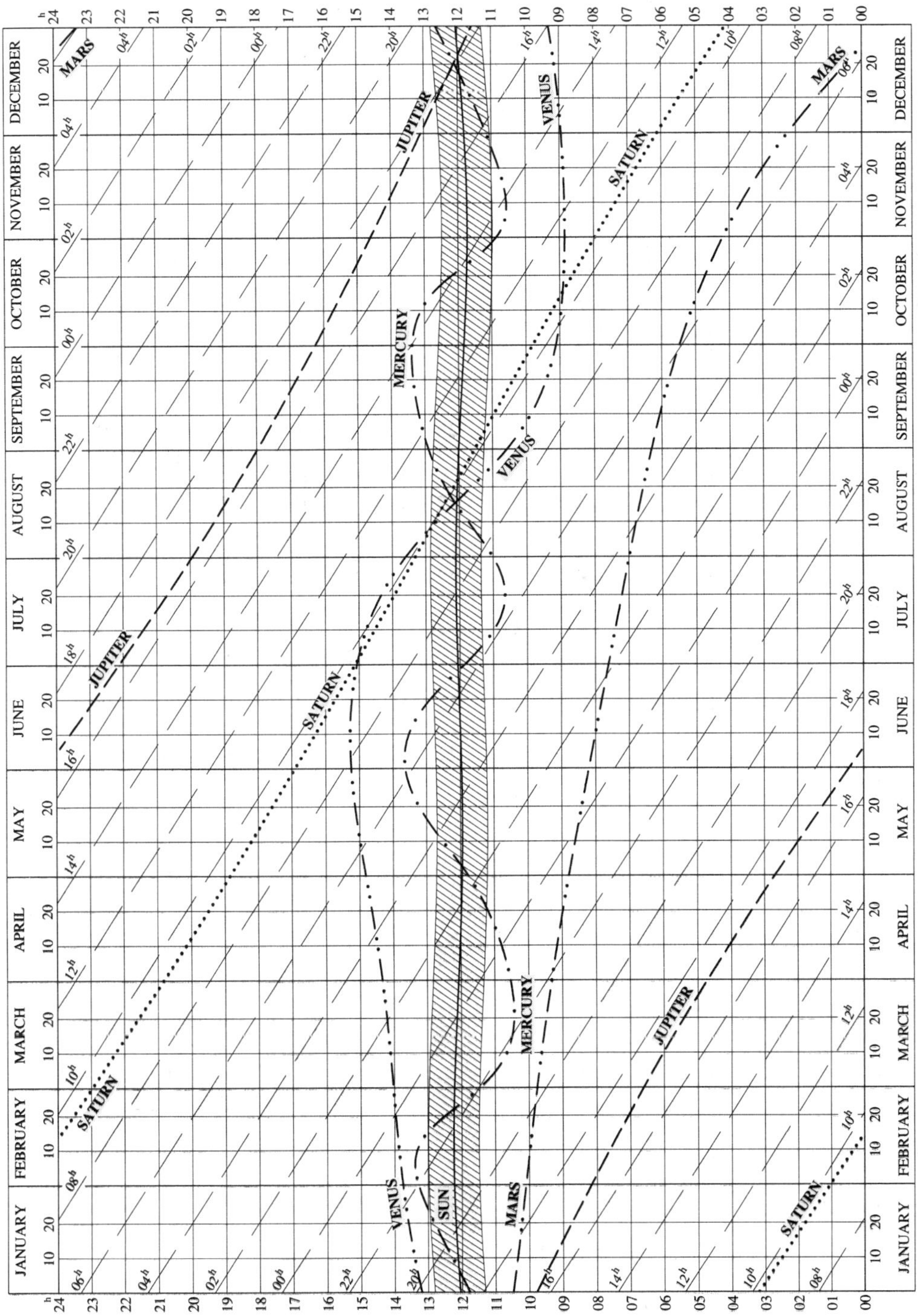

ELONGATIONS AND MAGNITUDES OF PLANETS AT 0^h UT

Date	Mercury Elong.	Mercury Mag.	Venus Elong.	Venus Mag.
	°		°	
Jan. 0	W. 4	−1·0	E. 16	−3·7
5	W. 2	−1·2	E. 17	−3·7
10	E. 3	−1·2	E. 18	−3·7
15	E. 5	−1·2	E. 19	−3·7
20	E. 9	−1·2	E. 20	−3·7
25	E. 12	−1·2	E. 22	−3·7
30	E. 15	−1·1	E. 23	−3·7
Feb. 4	E. 18	−0·9	E. 24	−3·7
9	E. 18	−0·4	E. 25	−3·7
14	E. 15	+0·4	E. 26	−3·7
19	E. 9	·	E. 27	−3·8
24	W. 4	·	E. 28	−3·8
Mar. 1	W. 12	·	E. 29	−3·8
6	W. 19	·	E. 30	−3·8
11	W. 24	+0·6	E. 32	−3·8
16	W. 27	+0·4	E. 33	−3·8
21	W. 28	+0·2	E. 34	−3·8
26	W. 27	+0·1	E. 35	−3·8
31	W. 26	0·0	E. 36	−3·8
Apr. 5	W. 24	−0·2	E. 37	−3·9
10	W. 21	−0·4	E. 38	−3·9
15	W. 18	−0·6	E. 39	−3·9
20	W. 14	−1·0	E. 40	−3·9
25	W. 9	−1·4	E. 41	−3·9
30	W. 4	−1·9	E. 41	−4·0
May 5	E. 2	−2·1	E. 42	−4·0
10	E. 8	−1·8	E. 43	−4·0
15	E. 14	−1·3	E. 44	−4·0
20	E. 18	−0·7	E. 44	−4·1
25	E. 21	−0·2	E. 45	−4·1
30	E. 23	+0·2	E. 45	−4·1
June 4	E. 23	+0·5	E. 45	−4·2
9	E. 22	+0·9	E. 45	−4·2
14	E. 19	·	E. 45	−4·3
19	E. 14	·	E. 45	−4·3
24	E. 8	·	E. 44	−4·3
29	W. 4	·	E. 44	−4·4
July 4	W. 9	·	E. 42	−4·4

Date	Mercury Elong.	Mercury Mag.	Venus Elong.	Venus Mag.
	°		°	
July 4	W. 9	·	E. 42	−4·4
9	W. 15	·	E. 41	−4·4
14	W. 18	·	E. 38	−4·4
19	W. 20	+0·4	E. 36	−4·4
24	W. 20	−0·2	E. 32	−4·4
29	W. 18	−0·7	E. 28	−4·3
Aug. 3	W. 14	−1·3	E. 22	−4·2
8	W. 9	−1·7	E. 17	−4·0
13	W. 3	−1·9	E. 11	−3·8
18	E. 3	−1·8	E. 8	−3·6
23	E. 7	−1·3	W. 11	−3·8
28	E. 11	−1·0	W. 17	−4·0
Sept. 2	E. 15	−0·7	W. 23	−4·2
7	E. 18	−0·5	W. 28	−4·4
12	E. 21	−0·3	W. 33	−4·5
17	E. 23	−0·2	W. 36	−4·5
22	E. 25	−0·1	W. 39	−4·5
27	E. 26	0·0	W. 41	−4·5
Oct. 2	E. 26	+0·1	W. 43	−4·5
7	E. 25	+0·2	W. 44	−4·5
12	E. 21	+0·5	W. 45	−4·4
17	E. 15	·	W. 46	−4·4
22	E. 5	·	W. 46	−4·4
27	W. 7	·	W. 46	−4·3
Nov. 1	W. 15	·	W. 46	−4·3
6	W. 19	−0·3	W. 46	−4·2
11	W. 19	−0·7	W. 46	−4·2
16	W. 17	−0·9	W. 46	−4·2
21	W. 15	−0·9	W. 45	−4·1
26	W. 12	−0·9	W. 44	−4·1
Dec. 1	W. 9	−1·0	W. 44	−4·1
6	W. 6	−1·0	W. 43	−4·0
11	W. 4	−1·1	W. 42	−4·0
16	W. 2	−1·2	W. 41	−4·0
21	E. 2	−1·1	W. 41	−4·0
26	E. 5	−1·1	W. 40	−3·9
31	E. 8	−1·1	W. 39	−3·9
36	E. 11	−1·1	W. 38	−3·9

MINOR PLANETS

	Conjunction	Stationary	Opposition	Stationary
Ceres	Mar. 22	Sept. 21	Nov. 9	—
Pallas	Feb. 1	July 6	Sept. 3	Oct. 22
Juno	Nov. 14	Feb. 19	Apr. 10	June 5
Vesta	—	Apr. 18	May 30	July 14

ELONGATIONS AND MAGNITUDES OF PLANETS AT 0^h UT

Date		Mars		Jupiter		Saturn		Uranus	Neptune	Pluto
		Elong.	Mag.	Elong.	Mag.	Elong.	Mag.	Elong.	Elong.	Elong.
		°		°		°		°	°	°
Jan.	**10**	W. 18	+1·5	W. 23	−1·7	W. 124	+0·3	E. 72	E. 49	W. 7
	0	W. 21	+1·5	W. 31	−1·8	W. 135	+0·3	E. 62	E. 39	W. 14
	10	W. 24	+1·5	W. 39	−1·8	W. 145	+0·2	E. 53	E. 29	W. 23
	20	W. 27	+1·4	W. 48	−1·8	W. 156	+0·1	E. 43	E. 19	W. 33
	30	W. 30	+1·4	W. 56	−1·9	W. 167	0·0	E. 33	E. 9	W. 42
Feb.	**9**	W. 33	+1·4	W. 65	−1·9	W. 178	0·0	E. 23	0	W. 52
	19	W. 35	+1·3	W. 73	−2·0	E. 171	0·0	E. 14	W. 10	W. 62
Mar.	**1**	W. 38	+1·3	W. 82	−2·0	E. 160	0·0	E. 4	W. 20	W. 71
	11	W. 40	+1·2	W. 91	−2·1	E. 149	+0·1	W. 5	W. 29	W. 81
	21	W. 43	+1·2	W. 101	−2·2	E. 139	+0·2	W. 14	W. 39	W. 91
	31	W. 45	+1·1	W. 110	−2·3	E. 129	+0·2	W. 24	W. 49	W. 101
Apr.	**10**	W. 47	+1·1	W. 120	−2·3	E. 118	+0·3	W. 33	W. 58	W. 111
	20	W. 49	+1·0	W. 130	−2·4	E. 109	+0·3	W. 42	W. 68	W. 120
	30	W. 51	+1·0	W. 140	−2·5	E. 99	+0·4	W. 52	W. 77	W. 130
May	**10**	W. 53	+1·0	W. 151	−2·5	E. 90	+0·4	W. 61	W. 87	W. 140
	20	W. 55	+0·9	W. 162	−2·6	E. 80	+0·5	W. 70	W. 97	W. 149
	30	W. 57	+0·9	W. 172	−2·6	E. 71	+0·5	W. 80	W. 106	W. 159
June	**9**	W. 59	+0·8	E. 177	−2·6	E. 62	+0·5	W. 89	W. 116	W. 168
	19	W. 62	+0·8	E. 166	−2·6	E. 54	+0·6	W. 99	W. 125	W. 173
	29	W. 64	+0·7	E. 155	−2·5	E. 45	+0·6	W. 108	W. 135	E. 168
July	**9**	W. 66	+0·7	E. 145	−2·5	E. 37	+0·6	W. 118	W. 145	E. 160
	19	W. 69	+0·6	E. 134	−2·5	E. 28	+0·6	W. 128	W. 155	E. 150
	29	W. 72	+0·5	E. 125	−2·4	E. 20	+0·6	W. 137	W. 164	E. 141
Aug.	**8**	W. 75	+0·5	E. 115	−2·3	E. 12	+0·6	W. 147	W. 174	E. 131
	18	W. 78	+0·4	E. 105	−2·3	E. 4	+0·6	W. 157	E. 176	E. 121
	28	W. 81	+0·3	E. 96	−2·2	W. 5	+0·6	W. 167	E. 166	E. 112
Sept.	**7**	W. 85	+0·2	E. 87	−2·1	W. 14	+0·6	W. 177	E. 156	E. 102
	17	W. 90	+0·1	E. 79	−2·1	W. 22	+0·7	E. 173	E. 146	E. 93
	27	W. 94	0·0	E. 70	−2·0	W. 31	+0·7	E. 163	E. 136	E. 83
Oct.	**7**	W. 100	−0·2	E. 62	−2·0	W. 39	+0·8	E. 152	E. 126	E. 73
	17	W. 106	−0·3	E. 54	−1·9	W. 48	+0·8	E. 142	E. 116	E. 64
	27	W. 113	−0·5	E. 45	−1·9	W. 57	+0·8	E. 132	E. 106	E. 54
Nov.	**6**	W. 121	−0·7	E. 37	−1·9	W. 66	+0·8	E. 122	E. 96	E. 44
	16	W. 131	−1·0	E. 29	−1·8	W. 76	+0·8	E. 112	E. 86	E. 35
	26	W. 142	−1·2	E. 22	−1·8	W. 85	+0·8	E. 101	E. 76	E. 25
Dec.	**6**	W. 154	−1·4	E. 14	−1·8	W. 95	+0·7	E. 91	E. 66	E. 16
	16	W. 167	−1·6	E. 6	−1·8	W. 105	+0·7	E. 81	E. 56	E. 8
	26	E. 176	−1·6	W. 2	−1·8	W. 115	+0·6	E. 71	E. 46	W. 8
	36	E. 164	−1·4	W. 10	−1·8	W. 126	+0·6	E. 61	E. 36	W. 16

Magnitudes at opposition: Uranus 5·7 Neptune 7·8 Pluto 13·9

VISUAL MAGNITUDES OF MINOR PLANETS

	Jan. 10	Feb. 19	Mar. 31	May 10	June 19	July 29	Sept. 7	Oct. 17	Nov. 26	Dec. 36
Ceres	9·3	9·2	9·0	9·2	9·2	9·0	8·5	7·7	7·5	8·3
Pallas	10·5	10·5	10·6	10·5	10·1	9·5	8·8	9·5	10·0	10·1
Juno	10·8	10·3	9·8	10·2	10·9	11·3	11·5	11·5	11·4	11·5
Vesta	7·8	7·4	6·7	5·8	5·8	6·6	7·3	7·7	7·9	8·0

CONFIGURATIONS OF SUN, MOON AND PLANETS

	d h		
Jan.	3 14	FULL MOON	
	3 20	Earth at perihelion	
	5 04	Jupiter 5° N. of *Antares*	
	6 18	Saturn 0°9 S. of Moon	Occn.
	7 05	*Regulus* 1°2 S. of Moon	Occn.
	7 06	Mercury in superior conjunction	
	10 16	Moon at apogee	
	11 13	LAST QUARTER	
	11 20	*Spica* 1°1 N. of Moon	Occn.
	15 13	*Antares* 0°5 N. of Moon	Occn.
	15 17	Jupiter 6° N. of Moon	
	17 02	Mars 5° N. of Moon	
	18 18	Venus 1°4 S. of Neptune	
	19 04	NEW MOON	
	20 13	Neptune 2° N. of Moon	
	20 17	Venus 0°8 N. of Moon	Occn.
	22 06	Uranus 0°4 S. of Moon	Occn.
	22 13	Moon at perigee	
	25 23	FIRST QUARTER	
Feb.	1 22	Pallas in conjunction with Sun	
	2 06	FULL MOON	
	2 23	Saturn 0°9 S. of Moon	Occn.
	3 14	*Regulus* 1°1 S. of Moon	Occn.
	7 13	Moon at apogee	
	7 13	Venus 0°7 S. of Uranus	
	7 17	Mercury greatest elong. E. (18°)	
	8 04	*Spica* 1°3 N. of Moon	Occn.
	8 16	Neptune in conjunction with Sun	
	10 10	LAST QUARTER	
	10 19	Saturn at opposition	
	11 22	*Antares* 0°7 N. of Moon	Occn.
	12 10	Jupiter 6° N. of Moon	
	13 14	Mercury stationary	
	15 01	Mars 4° N. of Moon	
	17 16	NEW MOON	
	19 00	Juno stationary	
	19 10	Moon at perigee	
	19 17	Venus 2° S. of Moon	
	23 05	Mercury in inferior conjunction	
	24 08	FIRST QUARTER	
Mar.	2 02	Saturn 1°1 S. of Moon	Occn.
	2 21	*Regulus* 1°1 S. of Moon	Occn.
	3 23	FULL MOON	Eclipse
	5 16	Uranus in conjunction with Sun	
	7 04	Moon at apogee	
	7 10	Mercury stationary	
	11 06	*Antares* 0°7 N. of Moon	Occn.
	12 00	Jupiter 6° N. of Moon	
	12 04	LAST QUARTER	
Mar.	16 01	Mars 1°9 N. of Moon	
	16 13	Neptune 2° N. of Moon	
	17 03	Mercury 1°4 N. of Moon	Occn.
	19 03	NEW MOON	Eclipse
	19 19	Moon at perigee	
	21 00	Equinox	
	21 15	Venus 4° S. of Moon	
	22 02	Mercury greatest elong. W. (28°)	
	22 04	Ceres in conjunction with Sun	
	25 07	Mars 1°0 S. of Neptune	
	25 18	FIRST QUARTER	
	29 04	Saturn 1°2 S. of Moon	Occn.
	30 03	*Regulus* 1°1 S. of Moon	Occn.
Apr.	1 00	Pluto stationary	
	1 07	Mercury 1°6 S. of Uranus	
	2 17	FULL MOON	
	3 09	Moon at apogee	
	6 02	Jupiter stationary	
	7 13	*Antares* 0°6 N. of Moon	Occn.
	8 09	Jupiter 6° N. of Moon	
	10 02	Juno at opposition	
	10 18	LAST QUARTER	
	13 00	Neptune 2° N. of Moon	
	14 02	Mars 0°5 S. of Moon	Occn.
	14 20	Uranus 1°0 S. of Moon	Occn.
	16 11	Mercury 5° S. of Moon	
	17 06	Moon at perigee	
	17 12	NEW MOON	
	18 18	Vesta stationary	
	20 01	Saturn stationary	
	20 08	Venus 3° S. of Moon	
	21 12	Venus 7° N. of *Aldebaran*	
	24 07	FIRST QUARTER	
	25 10	Saturn 1°1 S. of Moon	Occn.
	26 09	*Regulus* 1°0 S. of Moon	Occn.
	28 19	Mars 0°7 S. of Uranus	
	30 11	Moon at apogee	
May	2 10	FULL MOON	
	3 04	Mercury in superior conjunction	
	4 18	*Antares* 0°5 N. of Moon	Occn.
	5 12	Jupiter 6° N. of Moon	
	10 04	LAST QUARTER	
	10 08	Neptune 1°8 N. of Moon	
	12 07	Uranus 1°3 S. of Moon	Occn.
	13 01	Mars 3° S. of Moon	
	15 15	Moon at perigee	
	16 19	NEW MOON	
	16 21	Mercury 7° N. of *Aldebaran*	
	18 00	Mercury 3° S. of Moon	
	20 01	Venus 1°7 S. of Moon	

CONFIGURATIONS OF SUN, MOON AND PLANETS

	d	h		
May	22	19	Saturn 0°.8 S. of Moon	Occn.
	23	16	*Regulus* 0°.7 S. of Moon	Occn.
	23	21	FIRST QUARTER	
	25	06	Neptune stationary	
	27	22	Moon at apogee	
	30	14	Vesta at opposition	
	30	19	Venus 4° S. of *Pollux*	
June	1	01	*Antares* 0°.4 N. of Moon	Occn.
	1	01	FULL MOON	
	1	12	Jupiter 6° N. of Moon	
	2	10	Mercury greatest elong. E. (23°)	
	5	12	Juno stationary	
	5	23	Jupiter at opposition	
	6	14	Neptune 1°.5 N. of Moon	
	8	12	LAST QUARTER	
	8	15	Uranus 1°.6 S. of Moon	
	9	03	Venus greatest elong. E. (45°)	
	10	22	Mars 5° S. of Moon	
	12	17	Moon at perigee	
	15	03	NEW MOON	
	15	16	Mercury stationary	
	16	09	Mercury 6° S. of Moon	
	18	15	Venus 0°.6 S. of Moon	Occn.
	19	07	Pluto at opposition	
	19	08	Saturn 0°.4 S. of Moon	Occn.
	20	00	*Regulus* 0°.4 S. of Moon	Occn.
	21	18	Solstice	
	22	13	FIRST QUARTER	
	23	23	Uranus stationary	
	24	14	Moon at apogee	
	28	08	*Antares* 0°.5 N. of Moon	Occn.
	28	14	Jupiter 6° N. of Moon	
	28	19	Mercury in inferior conjunction	
	30	14	FULL MOON	
July	2	01	Venus 0°.8 S. of Saturn	
	3	19	Neptune 1°.3 N. of Moon	Occn.
	5	21	Uranus 1°.9 S. of Moon	
	6	14	Pallas stationary	
	7	00	Earth at aphelion	
	7	17	LAST QUARTER	
	9	15	Mars 6° S. of Moon	
	9	22	Moon at perigee	
	10	02	Mercury stationary	
	12	14	Venus greatest illuminated extent	
	13	03	Mercury 9° S. of Moon	
	14	10	Vesta stationary	
	14	12	NEW MOON	
	16	15	Venus 2° S. of *Regulus*	
	16	23	Saturn 0°.04 N. of Moon	Occn.
	17	09	*Regulus* 0°.3 S. of Moon	Occn.
	17	10	Venus 3° S. of Moon	
	20	15	Mercury greatest elong. W. (20°)	
July	22	06	FIRST QUARTER	
	22	09	Moon at apogee	
	25	13	Venus stationary	
	25	16	*Antares* 0°.6 N. of Moon	Occn.
	25	18	Jupiter 6° N. of Moon	
	30	01	FULL MOON	
	31	01	Neptune 1°.3 N. of Moon	Occn.
Aug.	1	19	Mercury 6° S. of *Pollux*	
	2	02	Uranus 2° S. of Moon	
	3	04	Venus 6° S. of *Regulus*	
	4	00	Moon at perigee	
	5	21	LAST QUARTER	
	7	04	Mars 6° S. of Moon	
	7	06	Jupiter stationary	
	12	23	NEW MOON	
	13	18	Neptune at opposition	
	15	20	Mercury in superior conjunction	
	18	04	Venus in inferior conjunction	
	19	03	Moon at apogee	
	21	00	FIRST QUARTER	
	21	23	Saturn in conjunction with Sun	
	22	01	*Antares* 0°.7 N. of Moon	Occn.
	22	03	Jupiter 6° N. of Moon	
	24	00	Mars 5° N. of *Aldebaran*	
	27	09	Neptune 1°.4 N. of Moon	
	28	11	FULL MOON	Eclipse
	29	09	Uranus 2° S. of Moon	
	31	00	Moon at perigee	
Sept.	3	00	Pallas at opposition	
	4	03	LAST QUARTER	
	4	14	Mars 6° S. of Moon	
	7	14	Venus stationary	
	7	22	Pluto stationary	
	8	19	Venus 9° S. of Moon	
	9	19	Uranus at opposition	
	10	01	*Regulus* 0°.2 S. of Moon	Occn.
	10	04	Saturn 0°.8 N. of Moon	Occn.
	11	13	NEW MOON	Eclipse
	13	14	Mercury 2° N. of Moon	
	15	21	Moon at apogee	
	18	08	*Antares* 0°.7 N. of Moon	Occn.
	18	15	Jupiter 6° N. of Moon	
	19	17	FIRST QUARTER	
	21	03	Ceres stationary	
	22	09	Mercury 0°.09 N. of *Spica*	
	23	10	Equinox	
	23	18	Neptune 1°.4 N. of Moon	
	23	23	Venus greatest illuminated extent	
	25	17	Uranus 1°.9 S. of Moon	

CONFIGURATIONS OF SUN, MOON AND PLANETS

	d h		
Sept.	26 20	FULL MOON	
	28 02	Moon at perigee	
	29 16	Mercury greatest elong. E. (26°)	
Oct.	2 20	Mars 5° S. of Moon	
	3 10	LAST QUARTER	
	7 03	Venus 3° S. of Moon	
	7 07	*Regulus* 0°.2 S. of Moon	Occn.
	7 16	Saturn 1°.3 N. of Moon	Occn.
	9 11	Venus 3° S. of *Regulus*	
	11 05	NEW MOON	
	12 07	Mercury stationary	
	13 01	Mercury 1°.3 N. of Moon	
	13 10	Moon at apogee	
	15 14	Venus 3° S. of Saturn	
	15 15	*Antares* 0°.5 N. of Moon	Occn.
	16 06	Jupiter 5° N. of Moon	
	19 09	FIRST QUARTER	
	21 03	Neptune 1°.3 N. of Moon	Occn.
	22 15	Pallas stationary	
	23 02	Uranus 1°.8 S. of Moon	
	24 00	Mercury in inferior conjunction	
	26 05	FULL MOON	
	26 12	Moon at perigee	
	28 15	Venus greatest elong. W. (46°)	
	30 19	Mars 3° S. of Moon	
	31 20	Neptune stationary	
Nov.	1 13	Mercury stationary	
	1 21	LAST QUARTER	
	3 13	*Regulus* 0°.03 N. of Moon	Occn.
	4 03	Saturn 1°.8 N. of Moon	
	5 20	Venus 3° N. of Moon	
	8 11	Mercury 7° N. of Moon	
	8 21	Mercury greatest elong. W. (19°)	
	9 13	Moon at apogee	
	9 15	Ceres at opposition	
	9 23	NEW MOON	
	11 21	*Antares* 0°.4 N. of Moon	Occn.
Nov.	12 22	Jupiter 5° N. of Moon	
	14 12	Juno in conjunction with Sun	
	15 16	Mars stationary	
	17 11	Neptune 1°.0 N. of Moon	Occn.
	17 23	FIRST QUARTER	
	19 11	Uranus 2° S. of Moon	
	24 00	Moon at perigee	
	24 14	FULL MOON	
	24 18	Uranus stationary	
	27 06	Mars 1°.7 S. of Moon	
	28 22	Venus 4° N. of *Spica*	
	30 20	*Regulus* 0°.3 N. of Moon	Occn.
Dec.	1 13	LAST QUARTER	
	1 13	Saturn 2° N. of Moon	
	6 01	Venus 7° N. of Moon	
	6 17	Moon at apogee	
	9 18	NEW MOON	
	12 21	Vesta 0°.4 N. of Moon	Occn.
	14 18	Neptune 0°.7 N. of Moon	Occn.
	16 01	Pallas 1°.0 S. of Moon	Occn.
	16 18	Uranus 2° S. of Moon	
	17 10	FIRST QUARTER	
	17 15	Mercury in superior conjunction	
	19 00	Mars closest approach	
	20 12	Saturn stationary	
	21 00	Pluto in conjunction with Sun	
	22 06	Solstice	
	22 10	Moon at perigee	
	23 06	Jupiter in conjunction with Sun	
	24 01	FULL MOON	
	24 03	Mars 0°.9 S. of Moon	Occn.
	24 20	Mars at opposition	
	28 05	*Regulus* 0°.6 N. of Moon	Occn.
	28 22	Saturn 3° N. of Moon	
	31 08	LAST QUARTER	

PREDICTED PERIHELION PASSAGES OF COMETS, 2007

Periodic comet	Perihelion date *T*	Perihelion distance *q* au	Period *P* years
99P/Kowal	Jan. 15	4·72	15·1
P/2001 WF_2 (LONEOS)	Feb. 6	0·99	5·0
P/2001 Q2 (Petriew)	Feb. 24	0·94	5·5
106P/Schuster	Apr. 2	1·56	7·3
96P/Machholz	Apr. 4	0·12	5·2
2P/Encke	Apr. 19	0·34	3·3
17P/Holmes	May 4	2·05	6·9
P/1998 QP_{54} (LONEOS-Tucker)	May 12	1·88	8·6
135P/Shoemaker-Levy	May 30	2·71	7·5
128P/Shoemaker-Holt	June 13	3·07	9·6
156P/Russell-LINEAR	June 17	1·59	6·8
87P/Bus	July 7	2·17	6·5
P/1998 U2 (Mueller)	July 7	2·03	8·7
108P/Ciffréo	July 18	1·72	7·3
P/2002 O5 (NEAT)	July 26	1·17	5·0
125P/Spacewatch	Aug. 10	1·52	5·5
70P/Kojima	Oct. 5	2·01	7·1
136P/Mueller	Oct. 22	2·96	8·6
50P/Arend	Nov. 1	1·92	8·3
P/1995 A1 (Jedicke)	Dec. 4	4·09	14·4
P/1990 V1 (Shoemaker-Levy)	Dec. 12	1·46	16·4
P/1998 S1 (LINEAR-Mueller)	Dec. 16	2·55	9·1
93P/Lovas	Dec. 17	1·70	9·2

CHRONOLOGICAL CYCLES AND ERAS

Dominical Letter	G	Julian Period (year of)	6720
Epact	11	Roman Indiction	15
Golden Number (Lunar Cycle)	XIII	Solar Cycle	28

All dates are given in terms of the Gregorian calendar in which 2007 January 14 corresponds to 2007 January 1 of the Julian calendar.

ERA	YEAR	BEGINS	ERA	YEAR	BEGINS
Byzantine	7516	Sept. 14	Japanese	2667	Jan. 1
Jewish (A.M.)*	5768	Sept. 12	Grecian (Seleucidæ)	2319	Sept. 14 (or Oct. 14)
Chinese (Ding-hai)		Feb. 18			
Roman (A.U.C.)	2760	Jan. 14	Indian (Saka)	1929	Mar. 22
Nabonassar	2756	Apr. 22	Diocletian	1724	Sept. 12
			Islamic (Hegira)*	1428	Jan. 19

* Year begins at sunset

RELIGIOUS CALENDARS

Epiphany	Jan. 6	Ascension Day	May 17
Ash Wednesday	Feb. 21	Whit Sunday—Pentecost	May 27
Palm Sunday	Apr. 1	Trinity Sunday	June 3
Good Friday	Apr. 6	First Sunday in Advent	Dec. 2
Easter Day	Apr. 8	Christmas Day (Tuesday)	Dec. 25
First Day of Passover (Pesach)	Apr. 3	Day of Atonement (Yom Kippur)	Sept. 22
Feast of Weeks (Shavuot)	May 23	First day of Tabernacles (Succoth)	Sept. 27
Jewish New Year (tabular) (Rosh Hashanah)	Sept. 13		
Islamic New Year (tabular)	Jan. 20	First day of Ramadân (tabular)	Sept. 13

The Jewish and Islamic dates above are tabular dates, which begin at sunset on the previous evening and end at sunset on the date tabulated. In practice, the dates of Islamic fasts and festivals are determined by an actual sighting of the appropriate new Moon.

CIVIL CALENDAR—UNITED STATES OF AMERICA

New Year's Day	Jan. 1	Labor Day	Sept. 3
Martin Luther King's Birthday	Jan. 15	Columbus Day	Oct. 8
Washington's Birthday	Feb. 19	Election Day (in certain States)	Nov. 6
Memorial Day	May 28	Veterans Day	Nov. 11
Independence Day	July 4	Thanksgiving Day	Nov. 22

CIVIL CALENDAR—UNITED KINGDOM

Accession of Queen Elizabeth II	Feb. 6	The Queen's Official Birthday†	June 9
St David (Wales)	Mar. 1	Birthday of Prince Philip, Duke of Edinburgh	June 10
Commonwealth Day	Mar. 12	Remembrance Sunday	Nov. 11
St Patrick (Ireland)	Mar. 17	Birthday of the Prince of Wales	Nov. 14
Birthday of Queen Elizabeth II	Apr. 21	St Andrew (Scotland)	Nov. 30
St George (England)	Apr. 23		
Coronation Day	June 2		

†Date subject to confirmation

Day of Month	JANUARY Day of Week	JANUARY Day of Year	FEBRUARY Day of Week	FEBRUARY Day of Year	MARCH Day of Week	MARCH Day of Year	APRIL Day of Week	APRIL Day of Year	MAY Day of Week	MAY Day of Year	JUNE Day of Week	JUNE Day of Year
1	Mon.	1	Thu.	32	Thu.	60	Sun.	91	Tue.	121	Fri.	152
2	Tue.	2	Fri.	33	Fri.	61	Mon.	92	Wed.	122	Sat.	153
3	Wed.	3	Sat.	34	Sat.	62	Tue.	93	Thu.	123	Sun.	154
4	Thu.	4	Sun.	35	Sun.	63	Wed.	94	Fri.	124	Mon.	155
5	Fri.	5	Mon.	36	Mon.	64	Thu.	95	Sat.	125	Tue.	156
6	Sat.	6	Tue.	37	Tue.	65	Fri.	96	Sun.	126	Wed.	157
7	Sun.	7	Wed.	38	Wed.	66	Sat.	97	Mon.	127	Thu.	158
8	Mon.	8	Thu.	39	Thu.	67	Sun.	98	Tue.	128	Fri.	159
9	Tue.	9	Fri.	40	Fri.	68	Mon.	99	Wed.	129	Sat.	160
10	Wed.	10	Sat.	41	Sat.	69	Tue.	100	Thu.	130	Sun.	161
11	Thu.	11	Sun.	42	Sun.	70	Wed.	101	Fri.	131	Mon.	162
12	Fri.	12	Mon.	43	Mon.	71	Thu.	102	Sat.	132	Tue.	163
13	Sat.	13	Tue.	44	Tue.	72	Fri.	103	Sun.	133	Wed.	164
14	Sun.	14	Wed.	45	Wed.	73	Sat.	104	Mon.	134	Thu.	165
15	Mon.	15	Thu.	46	Thu.	74	Sun.	105	Tue.	135	Fri.	166
16	Tue.	16	Fri.	47	Fri.	75	Mon.	106	Wed.	136	Sat.	167
17	Wed.	17	Sat.	48	Sat.	76	Tue.	107	Thu.	137	Sun.	168
18	Thu.	18	Sun.	49	Sun.	77	Wed.	108	Fri.	138	Mon.	169
19	Fri.	19	Mon.	50	Mon.	78	Thu.	109	Sat.	139	Tue.	170
20	Sat.	20	Tue.	51	Tue.	79	Fri.	110	Sun.	140	Wed.	171
21	Sun.	21	Wed.	52	Wed.	80	Sat.	111	Mon.	141	Thu.	172
22	Mon.	22	Thu.	53	Thu.	81	Sun.	112	Tue.	142	Fri.	173
23	Tue.	23	Fri.	54	Fri.	82	Mon.	113	Wed.	143	Sat.	174
24	Wed.	24	Sat.	55	Sat.	83	Tue.	114	Thu.	144	Sun.	175
25	Thu.	25	Sun.	56	Sun.	84	Wed.	115	Fri.	145	Mon.	176
26	Fri.	26	Mon.	57	Mon.	85	Thu.	116	Sat.	146	Tue.	177
27	Sat.	27	Tue.	58	Tue.	86	Fri.	117	Sun.	147	Wed.	178
28	Sun.	28	Wed.	59	Wed.	87	Sat.	118	Mon.	148	Thu.	179
29	Mon.	29			Thu.	88	Sun.	119	Tue.	149	Fri.	180
30	Tue.	30			Fri.	89	Mon.	120	Wed.	150	Sat.	181
31	Wed.	31			Sat.	90			Thu.	151		

JULIAN DATE, 2007

0^h UT	JD	0^h UT	JD	0^h UT	JD
Jan. 0	245 4100·5	May 0	245 4220·5	Sept. 0	245 4343·5
Feb. 0	245 4131·5	June 0	245 4251·5	Oct. 0	245 4373·5
Mar. 0	245 4159·5	July 0	245 4281·5	Nov. 0	245 4404·5
Apr. 0	245 4190·5	Aug. 0	245 4312·5	Dec. 0	245 4434·5

400-day date, JD 245 4400·5 = 2007 October 27·0

Standard epoch, 1900 January 0, 12^h UT = JD 241 5020·0
Standard epoch, B1950·0 = 1950 Jan. 0·923 = JD 243 3282·423
B2007·0 = 2007 Jan. 0·729 = JD 245 4101·229
J2007·5 = 2007 July 2·875 = JD 245 4284·375
Standard epoch, J2000·0 = 2000 Jan. 1·5 = JD 245 1545·0

	JULY		AUGUST		SEPTEMBER		OCTOBER		NOVEMBER		DECEMBER	
Day of Month	Day of Week	Day of Year	Day of Week	Day of Year	Day of Week	Day of Year	Day of Week	Day of Year	Day of Week	Day of Year	Day of Week	Day of Year
1	Sun.	182	Wed.	213	Sat.	244	Mon.	274	Thu.	305	Sat.	335
2	Mon.	183	Thu.	214	Sun.	245	Tue.	275	Fri.	306	Sun.	336
3	Tue.	184	Fri.	215	Mon.	246	Wed.	276	Sat.	307	Mon.	337
4	Wed.	185	Sat.	216	Tue.	247	Thu.	277	Sun.	308	Tue.	338
5	Thu.	186	Sun.	217	Wed.	248	Fri.	278	Mon.	309	Wed.	339
6	Fri.	187	Mon.	218	Thu.	249	Sat.	279	Tue.	310	Thu.	340
7	Sat.	188	Tue.	219	Fri.	250	Sun.	280	Wed.	311	Fri.	341
8	Sun.	189	Wed.	220	Sat.	251	Mon.	281	Thu.	312	Sat.	342
9	Mon.	190	Thu.	221	Sun.	252	Tue.	282	Fri.	313	Sun.	343
10	Tue.	191	Fri.	222	Mon.	253	Wed.	283	Sat.	314	Mon.	344
11	Wed.	192	Sat.	223	Tue.	254	Thu.	284	Sun.	315	Tue.	345
12	Thu.	193	Sun.	224	Wed.	255	Fri.	285	Mon.	316	Wed.	346
13	Fri.	194	Mon.	225	Thu.	256	Sat.	286	Tue.	317	Thu.	347
14	Sat.	195	Tue.	226	Fri.	257	Sun.	287	Wed.	318	Fri.	348
15	Sun.	196	Wed.	227	Sat.	258	Mon.	288	Thu.	319	Sat.	349
16	Mon.	197	Thu.	228	Sun.	259	Tue.	289	Fri.	320	Sun.	350
17	Tue.	198	Fri.	229	Mon.	260	Wed.	290	Sat.	321	Mon.	351
18	Wed.	199	Sat.	230	Tue.	261	Thu.	291	Sun.	322	Tue.	352
19	Thu.	200	Sun.	231	Wed.	262	Fri.	292	Mon.	323	Wed.	353
20	Fri.	201	Mon.	232	Thu.	263	Sat.	293	Tue.	324	Thu.	354
21	Sat.	202	Tue.	233	Fri.	264	Sun.	294	Wed.	325	Fri.	355
22	Sun.	203	Wed.	234	Sat.	265	Mon.	295	Thu.	326	Sat.	356
23	Mon.	204	Thu.	235	Sun.	266	Tue.	296	Fri.	327	Sun.	357
24	Tue.	205	Fri.	236	Mon.	267	Wed.	297	Sat.	328	Mon.	358
25	Wed.	206	Sat.	237	Tue.	268	Thu.	298	Sun.	329	Tue.	359
26	Thu.	207	Sun.	238	Wed.	269	Fri.	299	Mon.	330	Wed.	360
27	Fri.	208	Mon.	239	Thu.	270	Sat.	300	Tue.	331	Thu.	361
28	Sat.	209	Tue.	240	Fri.	271	Sun.	301	Wed.	332	Fri.	362
29	Sun.	210	Wed.	241	Sat.	272	Mon.	302	Thu.	333	Sat.	363
30	Mon.	211	Thu.	242	Sun.	273	Tue.	303	Fri.	334	Sun.	364
31	Tue.	212	Fri.	243			Wed.	304			Mon.	365

MEAN SIDEREAL TIME, 2007

Greenwich mean sidereal time at 0^h UT

	h		h		h		h
Jan. 0	6·6188	Apr. 0	12·5327	July 0	18·5123	Oct. 0	0·5576
Feb. 0	8·6558	May 0	14·5040	Aug. 0	20·5493	Nov. 0	2·5946
Mar. 0	10·4957	June 0	16·5410	Sept. 0	22·5863	Dec. 0	4·5659

Greenwich mean sidereal time (GMST) on day d of month at hour t UT

$$= \text{GMST at } 0^h \text{ UT on day } 0 + 0^h\!\!.065\,71\, d + 1^h\!\!.002\,74\, t$$

$$\text{Local mean sidereal time} = \text{GMST} \begin{matrix} + \text{ east} \\ - \text{ west} \end{matrix} \text{ longitude}$$

AT 0^h UNIVERSAL TIME

Date	Equation of time	Declin-ation	Date	Equation of time	Declin-ation	Date	Equation of time	Declin-ation	Date	Equation of time	Declin-ation
	m s	° ′		m s	° ′		m s	° ′		m s	° ′
Jan. 0	−02 43	−23 07	**Feb. 15**	−14 10	−12 53	**Apr. 1**	−04 08	+04 19	**May 17**	+03 38	+19 12
1	03 12	23 03	**16**	14 07	12 32	**2**	03 50	04 42	**18**	03 37	19 25
2	03 40	22 58	**17**	14 04	12 12	**3**	03 32	05 05	**19**	03 34	19 38
3	04 08	22 53	**18**	13 59	11 51	**4**	03 14	05 28	**20**	03 31	19 51
4	04 35	22 47	**19**	13 55	11 29	**5**	02 57	05 51	**21**	03 28	20 04
5	−05 03	−22 41	**20**	−13 49	−11 08	**6**	−02 39	+06 13	**22**	+03 24	+20 16
6	05 29	22 34	**21**	13 43	10 46	**7**	02 22	06 36	**23**	03 20	20 28
7	05 56	22 27	**22**	13 36	10 25	**8**	02 05	06 59	**24**	03 15	20 39
8	06 22	22 19	**23**	13 28	10 03	**9**	01 49	07 21	**25**	03 09	20 51
9	06 47	22 11	**24**	13 20	09 41	**10**	01 32	07 44	**26**	03 03	21 01
10	−07 12	−22 03	**25**	−13 12	−09 19	**11**	−01 16	+08 06	**27**	+02 57	+21 12
11	07 36	21 54	**26**	13 02	08 56	**12**	01 00	08 28	**28**	02 50	21 22
12	08 00	21 44	**27**	12 52	08 34	**13**	00 45	08 50	**29**	02 43	21 32
13	08 23	21 35	**28**	12 42	08 11	**14**	00 29	09 11	**30**	02 35	21 41
14	08 46	21 25	**Mar. 1**	12 31	07 49	**15**	−00 14	09 33	**31**	02 27	21 50
15	−09 08	−21 14	**2**	−12 19	−07 26	**16**	00 00	+09 55	**June 1**	+02 18	+21 58
16	09 29	21 03	**3**	12 07	07 03	**17**	+00 14	10 16	**2**	02 09	22 07
17	09 50	20 52	**4**	11 54	06 40	**18**	00 28	10 37	**3**	02 00	22 14
18	10 10	20 40	**5**	11 41	06 17	**19**	00 41	10 58	**4**	01 50	22 22
19	10 29	20 28	**6**	11 28	05 54	**20**	00 54	11 19	**5**	01 40	22 29
20	−10 48	−20 15	**7**	−11 14	−05 31	**21**	+01 07	+11 39	**6**	+01 30	+22 35
21	11 05	20 02	**8**	10 59	05 07	**22**	01 19	12 00	**7**	01 19	22 42
22	11 22	19 49	**9**	10 45	04 44	**23**	01 31	12 20	**8**	01 08	22 47
23	11 39	19 35	**10**	10 30	04 20	**24**	01 42	12 40	**9**	00 56	22 53
24	11 54	19 21	**11**	10 14	03 57	**25**	01 53	13 00	**10**	00 44	22 58
25	−12 09	−19 07	**12**	−09 58	−03 33	**26**	+02 03	+13 19	**11**	+00 32	+23 03
26	12 23	18 52	**13**	09 42	03 10	**27**	02 13	13 39	**12**	00 20	23 07
27	12 36	18 37	**14**	09 26	02 46	**28**	02 23	13 58	**13**	+00 08	23 11
28	12 48	18 21	**15**	09 10	02 22	**29**	02 32	14 17	**14**	−00 05	23 14
29	12 59	18 05	**16**	08 53	01 59	**30**	02 40	14 35	**15**	00 18	23 17
30	−13 10	−17 49	**17**	−08 36	−01 35	**May 1**	+02 48	+14 54	**16**	−00 31	+23 20
31	13 20	17 33	**18**	08 19	01 11	**2**	02 56	15 12	**17**	00 44	23 22
Feb. 1	13 29	17 16	**19**	08 02	00 48	**3**	03 02	15 30	**18**	00 57	23 24
2	13 37	16 59	**20**	07 44	−00 24	**4**	03 09	15 48	**19**	01 10	23 25
3	13 44	16 42	**21**	07 26	00 00	**5**	03 15	16 05	**20**	01 23	23 26
4	−13 50	−16 24	**22**	−07 09	+00 24	**6**	+03 20	+16 22	**21**	−01 36	+23 26
5	13 56	16 06	**23**	06 51	00 47	**7**	03 24	16 39	**22**	01 49	23 26
6	14 01	15 48	**24**	06 33	01 11	**8**	03 28	16 56	**23**	02 02	23 26
7	14 05	15 30	**25**	06 15	01 35	**9**	03 32	17 12	**24**	02 15	23 25
8	14 08	15 11	**26**	05 57	01 58	**10**	03 35	17 28	**25**	02 28	23 24
9	−14 11	−14 52	**27**	−05 38	+02 22	**11**	+03 37	+17 44	**26**	−02 41	+23 23
10	14 12	14 33	**28**	05 20	02 45	**12**	03 39	17 59	**27**	02 53	23 21
11	14 13	14 13	**29**	05 02	03 09	**13**	03 40	18 14	**28**	03 06	23 18
12	14 14	13 53	**30**	04 44	03 32	**14**	03 40	18 29	**29**	03 18	23 16
13	14 13	13 34	**31**	04 26	03 55	**15**	03 40	18 44	**30**	03 30	23 12
14	−14 12	−13 13	**Apr. 1**	−04 08	+04 19	**16**	+03 40	+18 58	**July 1**	−03 42	+23 09
15	−14 10	−12 53	**2**	−03 50	+04 42	**17**	+03 38	+19 12	**2**	−03 53	+23 05

Equation of time = apparent time − mean time

AT 0^h UNIVERSAL TIME

Date	Equation of time	Declination	Date	Equation of time	Declination	Date	Equation of time	Declination	Date	Equation of time	Declination
	m s	° ′		m s	° ′		m s	° ′		m s	° ′
July 1	−03 42	+23 09	**Aug. 16**	−04 27	+13 55	**Oct. 1**	+10 05	−02 57	**Nov. 16**	+15 23	−18 35
2	03 53	23 05	**17**	04 15	13 37	**2**	10 24	03 20	**17**	15 12	18 50
3	04 04	23 00	**18**	04 02	13 17	**3**	10 43	03 44	**18**	15 00	19 05
4	04 15	22 56	**19**	03 49	12 58	**4**	11 02	04 07	**19**	14 48	19 19
5	04 26	22 50	**20**	03 35	12 38	**5**	11 21	04 30	**20**	14 34	19 33
6	−04 37	+22 45	**21**	−03 21	+12 19	**6**	+11 39	−04 53	**21**	+14 20	−19 47
7	04 47	22 39	**22**	03 06	11 59	**7**	11 56	05 16	**22**	14 06	20 00
8	04 56	22 32	**23**	02 51	11 39	**8**	12 14	05 39	**23**	13 50	20 13
9	05 06	22 26	**24**	02 35	11 18	**9**	12 31	06 02	**24**	13 34	20 25
10	05 15	22 19	**25**	02 19	10 58	**10**	12 47	06 25	**25**	13 16	20 38
11	−05 23	+22 11	**26**	−02 03	+10 37	**11**	+13 03	−06 48	**26**	+12 58	−20 49
12	05 32	22 03	**27**	01 46	10 16	**12**	13 18	07 10	**27**	12 40	21 01
13	05 39	21 55	**28**	01 28	09 55	**13**	13 33	07 33	**28**	12 20	21 12
14	05 47	21 46	**29**	01 11	09 34	**14**	13 48	07 55	**29**	12 00	21 22
15	05 53	21 37	**30**	00 53	09 13	**15**	14 02	08 18	**30**	11 39	21 33
16	−06 00	+21 28	**31**	−00 34	+08 51	**16**	+14 15	−08 40	**Dec. 1**	+11 18	−21 42
17	06 05	21 18	**Sept. 1**	−00 15	08 30	**17**	14 28	09 02	**2**	10 55	21 52
18	06 11	21 08	**2**	+00 04	08 08	**18**	14 40	09 24	**3**	10 32	22 01
19	06 15	20 57	**3**	00 23	07 46	**19**	14 52	09 46	**4**	10 09	22 09
20	06 19	20 46	**4**	00 42	07 24	**20**	15 03	10 07	**5**	09 45	22 17
21	−06 23	+20 35	**5**	+01 02	+07 02	**21**	+15 14	−10 29	**6**	+09 20	−22 25
22	06 26	20 24	**6**	01 22	06 40	**22**	15 24	10 50	**7**	08 55	22 32
23	06 28	20 12	**7**	01 42	06 17	**23**	15 33	11 11	**8**	08 29	22 39
24	06 30	19 59	**8**	02 03	05 55	**24**	15 42	11 32	**9**	08 02	22 45
25	06 31	19 47	**9**	02 23	05 32	**25**	15 49	11 53	**10**	07 36	22 51
26	−06 32	+19 34	**10**	+02 44	+05 10	**26**	+15 57	−12 14	**11**	+07 08	−22 57
27	06 32	19 21	**11**	03 05	04 47	**27**	16 03	12 35	**12**	06 41	23 02
28	06 31	19 07	**12**	03 26	04 24	**28**	16 09	12 55	**13**	06 13	23 06
29	06 30	18 53	**13**	03 47	04 01	**29**	16 14	13 15	**14**	05 44	23 10
30	06 28	18 39	**14**	04 08	03 38	**30**	16 18	13 35	**15**	05 16	23 14
31	−06 26	+18 25	**15**	+04 30	+03 15	**31**	+16 21	−13 55	**16**	+04 47	−23 17
Aug. 1	06 23	18 10	**16**	04 51	02 52	**Nov. 1**	16 24	14 14	**17**	04 18	23 20
2	06 19	17 55	**17**	05 12	02 29	**2**	16 26	14 33	**18**	03 48	23 22
3	06 15	17 39	**18**	05 34	02 06	**3**	16 27	14 52	**19**	03 19	23 24
4	06 10	17 24	**19**	05 55	01 43	**4**	16 27	15 11	**20**	02 49	23 25
5	−06 04	+17 08	**20**	+06 17	+01 20	**5**	+16 26	−15 30	**21**	+02 20	−23 26
6	05 58	16 52	**21**	06 38	00 56	**6**	16 25	15 48	**22**	01 50	23 26
7	05 52	16 35	**22**	06 59	00 33	**7**	16 22	16 06	**23**	01 20	23 26
8	05 45	16 18	**23**	07 20	+00 10	**8**	16 19	16 24	**24**	00 50	23 26
9	05 37	16 01	**24**	07 42	−00 14	**9**	16 15	16 41	**25**	+00 21	23 25
10	−05 29	+15 44	**25**	+08 03	−00 37	**10**	+16 10	−16 58	**26**	−00 09	−23 23
11	05 20	15 27	**26**	08 23	01 00	**11**	16 04	17 15	**27**	00 39	23 21
12	05 10	15 09	**27**	08 44	01 24	**12**	15 57	17 32	**28**	01 08	23 19
13	05 00	14 51	**28**	09 05	01 47	**13**	15 50	17 48	**29**	01 37	23 16
14	04 50	14 33	**29**	09 25	02 11	**14**	15 42	18 04	**30**	02 07	23 12
15	−04 39	+14 14	**30**	+09 45	−02 34	**15**	+15 33	−18 20	**31**	−02 36	−23 08
16	−04 27	+13 55	**Oct. 1**	+10 05	−02 57	**16**	+15 23	−18 35	**32**	−03 04	−23 04

$$\text{UT of transit} = 12^h \begin{smallmatrix} -\ \text{east} \\ +\ \text{west} \end{smallmatrix} \text{ longitude} - \text{equation of time}$$

AT 0^h UNIVERSAL TIME

Date	*Polaris* GHA	σ Oct GHA	Date	*Polaris* GHA	σ Oct GHA	Date	*Polaris* GHA	σ Oct GHA	Date	*Polaris* GHA	σ Oct GHA
	° ′	° ′		° ′	° ′		° ′	° ′		° ′	° ′
Jan. 0	59 10	140 43	**Feb. 15**	104 49	186 04	**Apr. 1**	149 25	230 16	**May 17**	194 45	275 23
1	60 09	141 42	**16**	105 48	187 03	**2**	150 24	231 15	**18**	195 44	276 22
2	61 09	142 41	**17**	106 48	188 02	**3**	151 24	232 14	**19**	196 43	277 21
3	62 08	143 41	**18**	107 48	189 01	**4**	152 23	233 13	**20**	197 42	278 20
4	63 08	144 40	**19**	108 47	190 00	**5**	153 22	234 12	**21**	198 41	279 18
5	64 07	145 39	**20**	109 47	190 59	**6**	154 21	235 11	**22**	199 40	280 17
6	65 07	146 38	**21**	110 46	191 58	**7**	155 21	236 10	**23**	200 39	281 16
7	66 06	147 37	**22**	111 46	192 57	**8**	156 20	237 09	**24**	201 38	282 15
8	67 06	148 37	**23**	112 45	193 56	**9**	157 19	238 08	**25**	202 37	283 14
9	68 05	149 36	**24**	113 45	194 55	**10**	158 18	239 06	**26**	203 36	284 12
10	69 05	150 35	**25**	114 44	195 54	**11**	159 18	240 05	**27**	204 34	285 11
11	70 04	151 34	**26**	115 44	196 53	**12**	160 17	241 04	**28**	205 33	286 10
12	71 04	152 34	**27**	116 43	197 52	**13**	161 16	242 03	**29**	206 32	287 09
13	72 03	153 33	**28**	117 43	198 51	**14**	162 16	243 02	**30**	207 31	288 08
14	73 03	154 32	**Mar. 1**	118 42	199 50	**15**	163 15	244 00	**31**	208 30	289 07
15	74 02	155 31	**2**	119 42	200 49	**16**	164 14	244 59	**June 1**	209 29	290 06
16	75 02	156 31	**3**	120 42	201 48	**17**	165 13	245 58	**2**	210 27	291 04
17	76 01	157 30	**4**	121 41	202 47	**18**	166 12	246 57	**3**	211 26	292 03
18	77 01	158 29	**5**	122 41	203 46	**19**	167 12	247 56	**4**	212 25	293 02
19	78 00	159 28	**6**	123 40	204 45	**20**	168 11	248 55	**5**	213 24	294 01
20	79 00	160 27	**7**	124 40	205 44	**21**	169 10	249 54	**6**	214 23	295 00
21	80 00	161 26	**8**	125 39	206 43	**22**	170 09	250 52	**7**	215 22	295 59
22	80 59	162 25	**9**	126 39	207 42	**23**	171 08	251 51	**8**	216 21	296 57
23	81 59	163 25	**10**	127 38	208 41	**24**	172 07	252 50	**9**	217 19	297 56
24	82 59	164 24	**11**	128 37	209 40	**25**	173 06	253 49	**10**	218 18	298 55
25	83 58	165 23	**12**	129 37	210 39	**26**	174 05	254 48	**11**	219 17	299 54
26	84 58	166 22	**13**	130 36	211 38	**27**	175 05	255 46	**12**	220 16	300 53
27	85 57	167 21	**14**	131 36	212 36	**28**	176 04	256 45	**13**	221 15	301 52
28	86 57	168 20	**15**	132 35	213 35	**29**	177 03	257 44	**14**	222 13	302 51
29	87 56	169 20	**16**	133 35	214 34	**30**	178 02	258 43	**15**	223 12	303 50
30	88 56	170 19	**17**	134 34	215 33	**May 1**	179 01	259 42	**16**	224 11	304 48
31	89 55	171 18	**18**	135 34	216 32	**2**	180 00	260 41	**17**	225 09	305 47
Feb. 1	90 55	172 17	**19**	136 33	217 31	**3**	180 59	261 40	**18**	226 08	306 46
2	91 54	173 16	**20**	137 33	218 30	**4**	181 58	262 38	**19**	227 07	307 45
3	92 54	174 15	**21**	138 32	219 29	**5**	182 57	263 37	**20**	228 06	308 44
4	93 54	175 14	**22**	139 31	220 28	**6**	183 56	264 36	**21**	229 05	309 43
5	94 53	176 13	**23**	140 31	221 27	**7**	184 55	265 35	**22**	230 03	310 41
6	95 53	177 12	**24**	141 30	222 26	**8**	185 54	266 34	**23**	231 02	311 40
7	96 52	178 11	**25**	142 29	223 24	**9**	186 53	267 32	**24**	232 01	312 39
8	97 52	179 10	**26**	143 29	224 23	**10**	187 52	268 31	**25**	233 00	313 38
9	98 52	180 10	**27**	144 28	225 22	**11**	188 51	269 30	**26**	233 58	314 37
10	99 51	181 09	**28**	145 27	226 21	**12**	189 50	270 29	**27**	234 57	315 36
11	100 51	182 08	**29**	146 27	227 20	**13**	190 49	271 28	**28**	235 56	316 35
12	101 50	183 07	**30**	147 26	228 19	**14**	191 48	272 27	**29**	236 54	317 34
13	102 50	184 06	**31**	148 26	229 18	**15**	192 47	273 25	**30**	237 53	318 33
14	103 49	185 05	**Apr. 1**	149 25	230 16	**16**	193 46	274 24	**July 1**	238 52	319 32
15	104 49	186 04	**2**	150 24	231 15	**17**	194 45	275 23	**2**	239 50	320 31

The dates between Jan. 0 and Dec. 32 below are the dates when p changes to the next value.

Polar Distance (p) *Polaris*: Jan. 0 42′ Dec. 32
σ Octantis: Jan. 0 64′ Mar. 17 65′ Aug. 3 64′ Dec. 32

AT 0^h UNIVERSAL TIME

Date	*Polaris* GHA	*σ* Oct GHA	Date	*Polaris* GHA	*σ* Oct GHA	Date	*Polaris* GHA	*σ* Oct GHA	Date	*Polaris* GHA	*σ* Oct GHA
	° ′	° ′		° ′	° ′		° ′	° ′		° ′	° ′
July 1	238 52	319 32	**Aug. 16**	283 51	4 46	**Oct. 1**	328 52	50 11	**Nov. 16**	14 05	95 44
2	239 50	320 31	**17**	284 49	5 45	**2**	329 51	51 11	**17**	15 04	96 43
3	240 49	321 29	**18**	285 48	6 44	**3**	330 50	52 10	**18**	16 03	97 43
4	241 48	322 28	**19**	286 47	7 44	**4**	331 49	53 09	**19**	17 03	98 42
5	242 47	323 27	**20**	287 45	8 43	**5**	332 47	54 09	**20**	18 02	99 41
6	243 45	324 26	**21**	288 44	9 42	**6**	333 46	55 08	**21**	19 01	100 41
7	244 44	325 25	**22**	289 43	10 41	**7**	334 45	56 07	**22**	20 00	101 40
8	245 43	326 24	**23**	290 41	11 40	**8**	335 44	57 07	**23**	20 59	102 40
9	246 42	327 23	**24**	291 40	12 39	**9**	336 43	58 06	**24**	21 58	103 39
10	247 40	328 22	**25**	292 39	13 39	**10**	337 42	59 06	**25**	22 58	104 39
11	248 39	329 21	**26**	293 37	14 38	**11**	338 41	60 05	**26**	23 57	105 38
12	249 37	330 20	**27**	294 36	15 37	**12**	339 40	61 04	**27**	24 56	106 37
13	250 36	331 19	**28**	295 35	16 36	**13**	340 39	62 04	**28**	25 55	107 37
14	251 35	332 18	**29**	296 34	17 35	**14**	341 38	63 03	**29**	26 54	108 36
15	252 33	333 17	**30**	297 32	18 34	**15**	342 36	64 03	**30**	27 54	109 35
16	253 32	334 16	**31**	298 31	19 34	**16**	343 35	65 02	**Dec. 1**	28 53	110 35
17	254 31	335 15	**Sept. 1**	299 30	20 33	**17**	344 34	66 01	**2**	29 52	111 34
18	255 29	336 14	**2**	300 28	21 32	**18**	345 33	67 01	**3**	30 52	112 34
19	256 28	337 13	**3**	301 27	22 31	**19**	346 32	68 00	**4**	31 51	113 33
20	257 27	338 12	**4**	302 26	23 31	**20**	347 31	69 00	**5**	32 50	114 32
21	258 26	339 11	**5**	303 24	24 30	**21**	348 30	69 59	**6**	33 49	115 32
22	259 24	340 10	**6**	304 23	25 29	**22**	349 29	70 58	**7**	34 49	116 31
23	260 23	341 09	**7**	305 22	26 28	**23**	350 28	71 58	**8**	35 48	117 31
24	261 22	342 08	**8**	306 20	27 27	**24**	351 27	72 57	**9**	36 47	118 30
25	262 20	343 07	**9**	307 19	28 27	**25**	352 26	73 57	**10**	37 46	119 29
26	263 19	344 06	**10**	308 18	29 26	**26**	353 25	74 56	**11**	38 46	120 29
27	264 17	345 05	**11**	309 17	30 25	**27**	354 24	75 56	**12**	39 45	121 28
28	265 16	346 04	**12**	310 16	31 24	**28**	355 23	76 55	**13**	40 44	122 27
29	266 15	347 03	**13**	311 14	32 24	**29**	356 22	77 54	**14**	41 44	123 27
30	267 13	348 02	**14**	312 13	33 23	**30**	357 21	78 54	**15**	42 43	124 26
31	268 12	349 01	**15**	313 12	34 22	**31**	358 20	79 53	**16**	43 43	125 25
Aug. 1	269 11	350 00	**16**	314 11	35 22	**Nov. 1**	359 19	80 53	**17**	44 42	126 25
2	270 09	350 59	**17**	315 09	36 21	**2**	0 18	81 52	**18**	45 42	127 24
3	271 08	351 58	**18**	316 08	37 20	**3**	1 17	82 51	**19**	46 41	128 24
4	272 07	352 57	**19**	317 07	38 20	**4**	2 16	83 51	**20**	47 40	129 23
5	273 06	353 56	**20**	318 05	39 19	**5**	3 15	84 50	**21**	48 40	130 22
6	274 04	354 55	**21**	319 04	40 18	**6**	4 14	85 50	**22**	49 39	131 22
7	275 03	355 54	**22**	320 03	41 17	**7**	5 13	86 49	**23**	50 38	132 21
8	276 01	356 53	**23**	321 02	42 17	**8**	6 12	87 49	**24**	51 38	133 20
9	277 00	357 53	**24**	322 01	43 16	**9**	7 11	88 48	**25**	52 37	134 19
10	277 58	358 52	**25**	323 00	44 15	**10**	8 10	89 47	**26**	53 36	135 19
11	278 57	359 51	**26**	323 58	45 15	**11**	9 09	90 47	**27**	54 36	136 18
12	279 56	0 50	**27**	324 57	46 14	**12**	10 08	91 46	**28**	55 35	137 17
13	280 55	1 49	**28**	325 56	47 13	**13**	11 08	92 46	**29**	56 35	138 16
14	281 53	2 48	**29**	326 55	48 13	**14**	12 07	93 45	**30**	57 34	139 16
15	282 52	3 47	**30**	327 54	49 12	**15**	13 06	94 45	**31**	58 34	140 15
16	283 51	4 46	**Oct. 1**	328 52	50 11	**16**	14 05	95 44	**32**	59 33	141 14

Form the quantities $C = p \cos(\text{local hour angle})$ and $S = p \sin(\text{local hour angle})$, then
Latitude $= h_0 - C + 0{\cdot}0087\, S^2 \tan h_0$,
Azimuth of *Polaris* $= -S/\cos h_0$ and Azimuth of σ Octantis $= 180° + S/\cos h_0$, where p and h_0 are in degrees and h_0 is the observed altitude corrected for atmospheric refraction and instrument error.

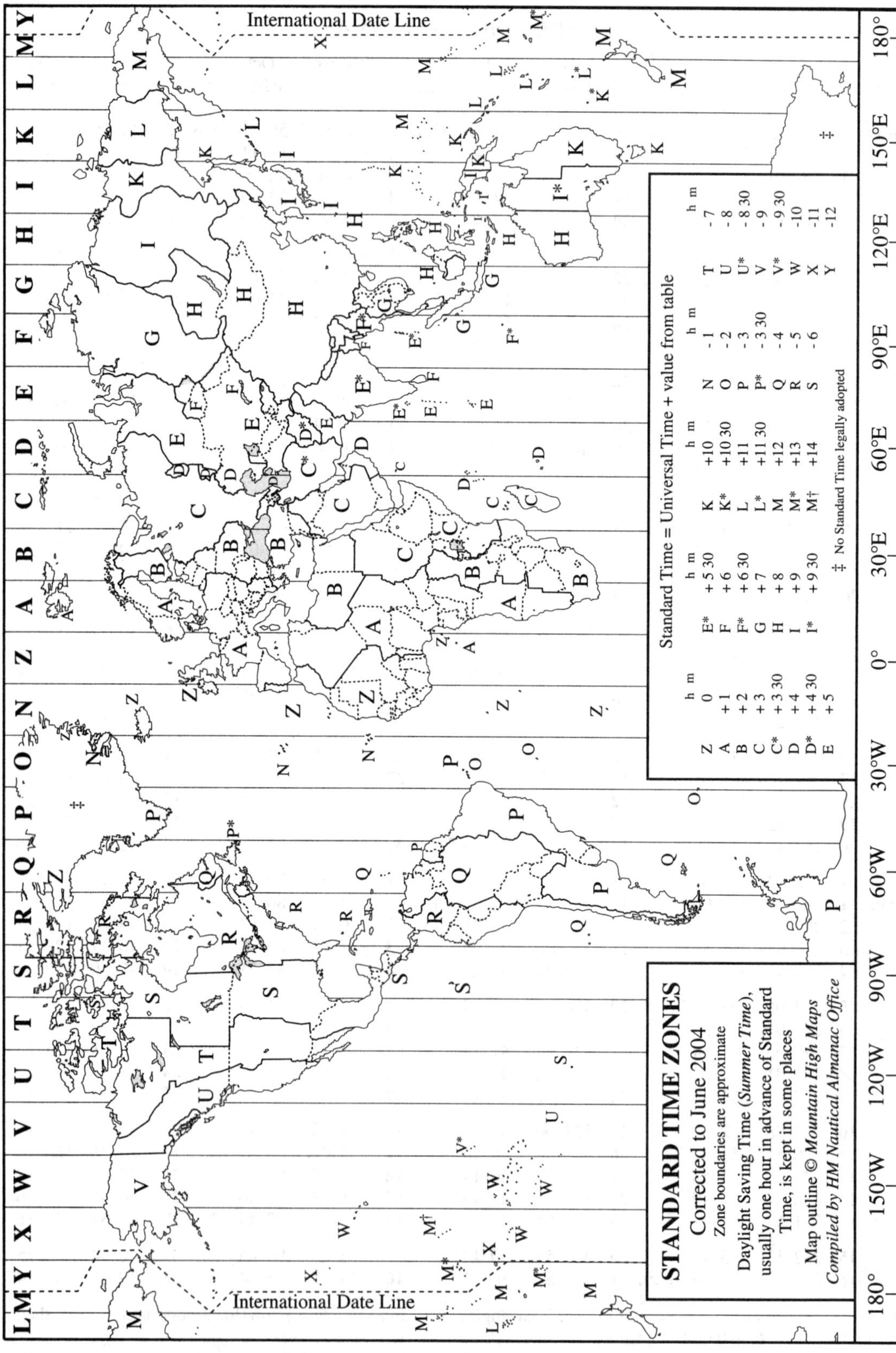
International Date Line
STANDARD TIME ZONES
Corrected to June 2004
Zone boundaries are approximate
Daylight Saving Time (Summer Time), usually one hour in advance of Standard Time, is kept in some places
Map outline © Mountain High Maps
Compiled by HM Nautical Almanac Office
Standard Time = Universal Time + value from table
Z 0
A +1
B +2
C +3
C* +3 30
D +4
D* +4 30
E +5
E* +5 30
F +6
F* +6 30
G +7
H +8
I +9
I* +9 30
K +10
K* +10 30
L +11
L* +11 30
M +12
M* +13
M† +14
N -1
O -2
P -3
P* -3 30
Q -4
R -5
S -6
T -7
U -8
U* -8 30
V -9
V* -9 30
W -10
X -11
Y -12
‡ No Standard Time legally adopted
180° 150°W 120°W 90°W 60°W 30°W 0° 30°E 60°E 90°E 120°E 150°E 180°

The times of sunrise and sunset (pages 24–31) and of moonrise and moonset (pages 32–63) are the instants when the upper limbs of the Sun and Moon appear to lie on the horizon for an observer at sea-level. In both cases a fixed allowance of 34′ has been made for refraction; a further allowance of 16′ has been made for the semidiameter of the Sun, while for the Moon the actual value of semidiameter *minus* horizontal parallax has been used. No allowance has been made for the phase of the Moon. The observed times may differ from the tabular times because of variations in refraction and the relative heights of the observer and horizon.

The tabular values are for the universal time (UT) of the phenomena on the Greenwich meridian (longitude 0°). To a first approximation the UT at another longitude is given by subtracting the longitude, expressed in time-measure, if east of Greenwich, or by adding, if west of Greenwich. Alternatively the tables may be regarded as giving the approximate local mean time on all meridians. These times may be converted to standard time by applying the appropriate differences, as indicated in the note on page 4. Linear interpolation may be used to obtain the times for non-tabular latitudes.

In the case of the Sun it may be necessary to interpolate (mentally) to obtain the UT for an intermediate date, but a further interpolation for longitude is not normally required. In the case of the Moon the values must normally be interpolated for longitude, as well as for latitude, since the changes in the tabular values from one day to the next are usually large. The interpolating factor is equal to one twenty-fourth of the longitude if expressed in hours and decimals of an hour; linear interpolation is usually adequate.

Example

To find the times of sunrise and sunset and of moonrise and moonset on 2007 February 4 at latitude N 38° 55′, longitude W 77° 15′. The longitude expressed in time-measure is W 05^h 09^m. The difference between standard time and UT is -5^h in this case.

The relative tabular values in UT for longitude 0° are as follows:

	Sunrise +35°	Sunrise +40°	Sunset +35°	Sunset +40°		Moonrise +35°	Moonrise +40°	Moonset +35°	Moonset +40°
d	h m	h m	h m	h m	d	h m	h m	h m	h m
Feb. 3	06 58	07 07	17 30	17 21	**Feb. 4**	19 49	19 46	08 11	08 16
7	06 54	07 03	17 34	17 26	5	20 46	20 46	08 35	08 36

Interpolating factor for latitude is 3° 55′/ 5° = 0·78
for date for Sun is $1^d/4^d$ = 0·25
for long. for Moon is $5^h.15/24^h$ = 0·21

		Sunrise	Sunset		Moonrise	Moonset
	d	h m	h m	d	h m	h m
Interpolation to:						
Latitude N 38° 55′	**Feb. 3**	07 05	17 23	**Feb. 4**	19 47	08 15
N 38° 55′	7	07 01	17 28	5	20 46	08 36
Local mean time	4	07 04	17 24	4	19 59	08 19
Adjustment to:						
Universal time	4	12 13	22 33	4	01 08	13 28
Standard time	4	07 13	17 33	4	20 08	08 28

UNIVERSAL TIME FOR MERIDIAN OF GREENWICH

SUNRISE

Lat.	−55°	−50°	−45°	−40°	−35°	−30°	−20°	−10°	0°	+10°	+20°	+30°	+35°	+40°
	h m	h m	h m	h m	h m	h m	h m	h m	h m	h m	h m	h m	h m	h m
Jan. −2	3 22	3 52	4 14	4 32	4 47	5 00	5 22	5 41	5 58	6 16	6 34	6 55	7 07	7 21
2	3 27	3 56	4 18	4 35	4 50	5 03	5 24	5 43	6 00	6 17	6 35	6 56	7 08	7 22
6	3 32	4 00	4 22	4 39	4 53	5 06	5 27	5 45	6 02	6 19	6 36	6 57	7 09	7 22
10	3 38	4 06	4 26	4 43	4 57	5 09	5 30	5 47	6 04	6 20	6 37	6 57	7 08	7 22
14	3 45	4 11	4 31	4 47	5 01	5 12	5 32	5 49	6 05	6 21	6 38	6 57	7 08	7 21
18	3 53	4 17	4 36	4 52	5 05	5 16	5 35	5 51	6 07	6 22	6 38	6 56	7 07	7 19
22	4 01	4 24	4 42	4 56	5 09	5 19	5 38	5 53	6 08	6 22	6 38	6 55	7 05	7 17
26	4 09	4 31	4 47	5 01	5 13	5 23	5 40	5 55	6 09	6 23	6 37	6 54	7 03	7 14
30	4 17	4 37	4 53	5 06	5 17	5 26	5 43	5 57	6 10	6 23	6 36	6 52	7 01	7 11
Feb. 3	4 26	4 44	4 59	5 11	5 21	5 30	5 45	5 58	6 10	6 22	6 35	6 49	6 58	7 07
7	4 35	4 52	5 05	5 16	5 25	5 33	5 47	5 59	6 11	6 22	6 33	6 47	6 54	7 03
11	4 43	4 59	5 11	5 21	5 29	5 37	5 50	6 01	6 11	6 21	6 32	6 44	6 51	6 58
15	4 52	5 06	5 17	5 26	5 33	5 40	5 52	6 01	6 11	6 20	6 29	6 40	6 47	6 54
19	5 01	5 13	5 22	5 30	5 37	5 43	5 53	6 02	6 10	6 19	6 27	6 37	6 42	6 48
23	5 09	5 20	5 28	5 35	5 41	5 46	5 55	6 03	6 10	6 17	6 24	6 33	6 37	6 43
27	5 18	5 27	5 34	5 40	5 45	5 49	5 57	6 03	6 09	6 15	6 22	6 29	6 33	6 37
Mar. 3	5 26	5 33	5 39	5 44	5 48	5 52	5 58	6 04	6 09	6 14	6 19	6 24	6 28	6 31
7	5 34	5 40	5 45	5 49	5 52	5 55	6 00	6 04	6 08	6 12	6 16	6 20	6 22	6 25
11	5 42	5 47	5 50	5 53	5 55	5 57	6 01	6 04	6 07	6 09	6 12	6 15	6 17	6 19
15	5 50	5 53	5 55	5 57	5 59	6 00	6 02	6 04	6 06	6 07	6 09	6 10	6 11	6 12
19	5 58	5 59	6 00	6 01	6 02	6 03	6 03	6 04	6 05	6 05	6 05	6 06	6 06	6 06
23	6 06	6 06	6 06	6 05	6 05	6 05	6 05	6 04	6 03	6 03	6 02	6 01	6 00	5 59
27	6 14	6 12	6 11	6 09	6 08	6 07	6 06	6 04	6 02	6 00	5 58	5 56	5 55	5 53
31	6 22	6 18	6 16	6 14	6 12	6 10	6 07	6 04	6 01	5 58	5 55	5 51	5 49	5 46
Apr. 4	6 29	6 25	6 21	6 18	6 15	6 12	6 08	6 04	6 00	5 56	5 51	5 46	5 43	5 40

SUNSET

Lat.	−55°	−50°	−45°	−40°	−35°	−30°	−20°	−10°	0°	+10°	+20°	+30°	+35°	+40°
	h m	h m	h m	h m	h m	h m	h m	h m	h m	h m	h m	h m	h m	h m
Jan. −2	20 41	20 12	19 49	19 32	19 17	19 04	18 42	18 23	18 06	17 48	17 30	17 09	16 57	16 43
2	20 40	20 11	19 50	19 32	19 17	19 05	18 43	18 25	18 08	17 51	17 33	17 12	17 00	16 46
6	20 39	20 10	19 49	19 32	19 18	19 05	18 44	18 26	18 09	17 53	17 35	17 15	17 03	16 50
10	20 36	20 09	19 48	19 31	19 18	19 06	18 45	18 27	18 11	17 55	17 38	17 18	17 07	16 53
14	20 32	20 06	19 46	19 30	19 17	19 05	18 45	18 28	18 13	17 57	17 40	17 21	17 10	16 58
18	20 27	20 02	19 44	19 28	19 16	19 05	18 46	18 29	18 14	17 59	17 43	17 25	17 14	17 02
22	20 21	19 58	19 41	19 26	19 14	19 03	18 45	18 30	18 15	18 01	17 46	17 28	17 18	17 07
26	20 15	19 53	19 37	19 23	19 12	19 02	18 45	18 30	18 16	18 02	17 48	17 32	17 22	17 11
30	20 08	19 48	19 33	19 20	19 09	19 00	18 44	18 30	18 17	18 04	17 50	17 35	17 26	17 16
Feb. 3	20 00	19 42	19 28	19 16	19 06	18 57	18 42	18 29	18 17	18 05	17 53	17 39	17 30	17 21
7	19 52	19 36	19 22	19 12	19 02	18 54	18 41	18 29	18 18	18 07	17 55	17 42	17 34	17 26
11	19 44	19 29	19 17	19 07	18 59	18 51	18 39	18 28	18 18	18 08	17 57	17 45	17 38	17 31
15	19 35	19 21	19 11	19 02	18 54	18 48	18 36	18 27	18 18	18 09	17 59	17 48	17 42	17 35
19	19 26	19 14	19 04	18 57	18 50	18 44	18 34	18 25	18 17	18 09	18 01	17 51	17 46	17 40
23	19 16	19 06	18 58	18 51	18 45	18 40	18 31	18 24	18 17	18 10	18 03	17 54	17 50	17 45
27	19 07	18 58	18 51	18 45	18 40	18 36	18 28	18 22	18 16	18 10	18 04	17 57	17 53	17 49
Mar. 3	18 57	18 50	18 44	18 39	18 35	18 32	18 25	18 20	18 15	18 11	18 06	18 00	17 57	17 54
7	18 47	18 41	18 37	18 33	18 30	18 27	18 22	18 18	18 14	18 11	18 07	18 03	18 00	17 58
11	18 37	18 33	18 29	18 27	18 24	18 22	18 19	18 16	18 13	18 11	18 08	18 05	18 04	18 02
15	18 27	18 24	18 22	18 20	18 19	18 18	18 16	18 14	18 12	18 11	18 09	18 08	18 07	18 06
19	18 16	18 15	18 14	18 14	18 13	18 13	18 12	18 12	18 11	18 11	18 11	18 11	18 10	18 10
23	18 06	18 07	18 07	18 07	18 08	18 08	18 09	18 09	18 10	18 11	18 12	18 13	18 14	18 15
27	17 56	17 58	17 59	18 01	18 02	18 03	18 05	18 07	18 09	18 11	18 13	18 15	18 17	18 19
31	17 46	17 49	17 52	17 54	17 56	17 58	18 02	18 05	18 08	18 11	18 14	18 18	18 20	18 23
Apr. 4	17 36	17 41	17 45	17 48	17 51	17 54	17 58	18 02	18 06	18 10	18 15	18 20	18 23	18 27

UNIVERSAL TIME FOR MERIDIAN OF GREENWICH

SUNRISE

Lat.	+40°	+42°	+44°	+46°	+48°	+50°	+52°	+54°	+56°	+58°	+60°	+62°	+64°	+66°
	h m	h m	h m	h m	h m	h m	h m	h m	h m	h m	h m	h m	h m	h m
Jan. −2	7 21	7 28	7 34	7 42	7 50	7 58	8 08	8 19	8 32	8 46	9 03	9 25	9 52	10 33
2	7 22	7 28	7 35	7 42	7 50	7 59	8 08	8 19	8 31	8 45	9 02	9 22	9 49	10 27
6	7 22	7 28	7 35	7 42	7 49	7 58	8 07	8 17	8 29	8 43	8 59	9 19	9 44	10 18
10	7 22	7 27	7 34	7 41	7 48	7 56	8 05	8 15	8 27	8 40	8 55	9 14	9 37	10 08
14	7 21	7 26	7 32	7 39	7 46	7 54	8 02	8 12	8 23	8 35	8 50	9 07	9 29	9 57
18	7 19	7 24	7 30	7 36	7 43	7 51	7 59	8 08	8 18	8 30	8 43	9 00	9 19	9 45
22	7 17	7 22	7 27	7 33	7 40	7 47	7 54	8 03	8 13	8 24	8 36	8 51	9 09	9 32
26	7 14	7 19	7 24	7 30	7 36	7 42	7 49	7 57	8 06	8 16	8 28	8 42	8 58	9 18
30	7 11	7 15	7 20	7 25	7 31	7 37	7 44	7 51	7 59	8 09	8 19	8 32	8 47	9 05
Feb. 3	7 07	7 11	7 16	7 20	7 26	7 31	7 37	7 44	7 52	8 00	8 10	8 21	8 35	8 50
7	7 03	7 07	7 11	7 15	7 20	7 25	7 31	7 37	7 44	7 51	8 00	8 10	8 22	8 36
11	6 58	7 02	7 06	7 10	7 14	7 18	7 23	7 29	7 35	7 42	7 50	7 59	8 09	8 22
15	6 54	6 57	7 00	7 03	7 07	7 11	7 16	7 21	7 26	7 32	7 39	7 47	7 56	8 07
19	6 48	6 51	6 54	6 57	7 00	7 04	7 08	7 12	7 17	7 22	7 28	7 35	7 43	7 52
23	6 43	6 45	6 48	6 50	6 53	6 56	7 00	7 03	7 07	7 12	7 17	7 23	7 29	7 37
27	6 37	6 39	6 41	6 43	6 46	6 48	6 51	6 54	6 58	7 01	7 06	7 10	7 16	7 22
Mar. 3	6 31	6 33	6 34	6 36	6 38	6 40	6 42	6 45	6 48	6 51	6 54	6 58	7 02	7 07
7	6 25	6 26	6 27	6 29	6 30	6 32	6 34	6 35	6 37	6 40	6 42	6 45	6 48	6 52
11	6 19	6 20	6 20	6 21	6 22	6 23	6 24	6 26	6 27	6 29	6 30	6 32	6 34	6 37
15	6 12	6 13	6 13	6 14	6 14	6 15	6 15	6 16	6 17	6 17	6 18	6 19	6 20	6 21
19	6 06	6 06	6 06	6 06	6 06	6 06	6 06	6 06	6 06	6 06	6 06	6 06	6 06	6 06
23	5 59	5 59	5 59	5 58	5 58	5 57	5 57	5 56	5 56	5 55	5 54	5 53	5 52	5 51
27	5 53	5 52	5 51	5 51	5 50	5 49	5 48	5 46	5 45	5 44	5 42	5 40	5 38	5 35
31	5 46	5 45	5 44	5 43	5 41	5 40	5 38	5 36	5 34	5 32	5 30	5 27	5 24	5 20
Apr. 4	5 40	5 39	5 37	5 35	5 33	5 31	5 29	5 27	5 24	5 21	5 18	5 14	5 09	5 04

SUNSET

Lat.	+40°	+42°	+44°	+46°	+48°	+50°	+52°	+54°	+56°	+58°	+60°	+62°	+64°	+66°
	h m	h m	h m	h m	h m	h m	h m	h m	h m	h m	h m	h m	h m	h m
Jan. −2	16 43	16 37	16 30	16 22	16 14	16 06	15 56	15 45	15 32	15 18	15 01	14 40	14 12	13 32
2	16 46	16 40	16 33	16 26	16 18	16 09	16 00	15 49	15 37	15 23	15 06	14 46	14 19	13 41
6	16 50	16 43	16 37	16 30	16 22	16 14	16 05	15 54	15 42	15 29	15 13	14 53	14 28	13 53
10	16 53	16 48	16 41	16 35	16 27	16 19	16 10	16 00	15 49	15 36	15 20	15 02	14 38	14 07
14	16 58	16 52	16 46	16 39	16 32	16 25	16 16	16 06	15 56	15 43	15 29	15 11	14 50	14 21
18	17 02	16 57	16 51	16 45	16 38	16 31	16 22	16 13	16 03	15 51	15 38	15 22	15 02	14 37
22	17 07	17 02	16 56	16 50	16 44	16 37	16 29	16 21	16 11	16 00	15 47	15 33	15 15	14 52
26	17 11	17 07	17 01	16 56	16 50	16 43	16 36	16 28	16 19	16 09	15 58	15 44	15 28	15 07
30	17 16	17 12	17 07	17 02	16 56	16 50	16 43	16 36	16 28	16 19	16 08	15 56	15 41	15 23
Feb. 3	17 21	17 17	17 12	17 08	17 03	16 57	16 51	16 44	16 37	16 28	16 18	16 07	15 54	15 38
7	17 26	17 22	17 18	17 14	17 09	17 04	16 58	16 52	16 45	16 38	16 29	16 19	16 07	15 53
11	17 31	17 27	17 23	17 20	17 15	17 11	17 06	17 00	16 54	16 47	16 40	16 31	16 20	16 08
15	17 35	17 32	17 29	17 25	17 22	17 18	17 13	17 08	17 03	16 57	16 50	16 42	16 33	16 23
19	17 40	17 37	17 34	17 31	17 28	17 25	17 21	17 16	17 12	17 06	17 01	16 54	16 46	16 37
23	17 45	17 42	17 40	17 37	17 34	17 31	17 28	17 24	17 20	17 16	17 11	17 05	16 59	16 51
27	17 49	17 47	17 45	17 43	17 41	17 38	17 35	17 32	17 29	17 25	17 21	17 17	17 11	17 05
Mar. 3	17 54	17 52	17 50	17 49	17 47	17 45	17 43	17 40	17 37	17 35	17 31	17 28	17 23	17 19
7	17 58	17 57	17 55	17 54	17 53	17 51	17 50	17 48	17 46	17 44	17 41	17 39	17 36	17 32
11	18 02	18 01	18 01	18 00	17 59	17 58	17 57	17 56	17 54	17 53	17 51	17 50	17 48	17 45
15	18 06	18 06	18 06	18 05	18 05	18 04	18 04	18 03	18 03	18 02	18 01	18 00	18 00	17 58
19	18 10	18 11	18 11	18 11	18 11	18 11	18 11	18 11	18 11	18 11	18 11	18 11	18 11	18 12
23	18 15	18 15	18 15	18 16	18 16	18 17	18 18	18 18	18 19	18 20	18 21	18 22	18 23	18 25
27	18 19	18 20	18 20	18 21	18 22	18 23	18 24	18 26	18 27	18 29	18 31	18 33	18 35	18 38
31	18 23	18 24	18 25	18 27	18 28	18 30	18 31	18 33	18 35	18 38	18 40	18 43	18 47	18 51
Apr. 4	18 27	18 28	18 30	18 32	18 34	18 36	18 38	18 41	18 43	18 47	18 50	18 54	18 59	19 04

UNIVERSAL TIME FOR MERIDIAN OF GREENWICH

SUNRISE

Lat.	−55°	−50°	−45°	−40°	−35°	−30°	−20°	−10°	0°	+10°	+20°	+30°	+35°	+40°
	h m	h m	h m	h m	h m	h m	h m	h m	h m	h m	h m	h m	h m	h m
Mar. 31	6 22	6 18	6 16	6 14	6 12	6 10	6 07	6 04	6 01	5 58	5 55	5 51	5 49	5 46
Apr. 4	6 29	6 25	6 21	6 18	6 15	6 12	6 08	6 04	6 00	5 56	5 51	5 46	5 43	5 40
8	6 37	6 31	6 26	6 22	6 18	6 15	6 09	6 04	5 59	5 54	5 48	5 42	5 38	5 34
12	6 45	6 37	6 31	6 26	6 21	6 17	6 10	6 04	5 58	5 51	5 45	5 37	5 33	5 28
16	6 52	6 43	6 36	6 30	6 24	6 19	6 11	6 04	5 57	5 49	5 42	5 33	5 27	5 21
20	7 00	6 49	6 41	6 34	6 27	6 22	6 12	6 04	5 56	5 47	5 39	5 28	5 22	5 16
24	7 07	6 55	6 46	6 38	6 31	6 24	6 14	6 04	5 55	5 46	5 36	5 24	5 18	5 10
28	7 15	7 01	6 51	6 41	6 34	6 27	6 15	6 04	5 54	5 44	5 33	5 20	5 13	5 05
May 2	7 22	7 07	6 55	6 45	6 37	6 29	6 16	6 05	5 54	5 42	5 31	5 17	5 09	4 59
6	7 30	7 13	7 00	6 49	6 40	6 32	6 18	6 05	5 53	5 41	5 28	5 13	5 05	4 55
10	7 37	7 19	7 05	6 53	6 43	6 34	6 19	6 06	5 53	5 40	5 26	5 10	5 01	4 50
14	7 44	7 25	7 10	6 57	6 46	6 37	6 21	6 06	5 53	5 39	5 25	5 08	4 58	4 46
18	7 51	7 30	7 14	7 01	6 49	6 39	6 22	6 07	5 53	5 39	5 23	5 05	4 55	4 43
22	7 57	7 35	7 18	7 04	6 52	6 42	6 24	6 08	5 53	5 38	5 22	5 03	4 52	4 39
26	8 03	7 40	7 22	7 08	6 55	6 44	6 25	6 09	5 53	5 38	5 21	5 01	4 50	4 37
30	8 09	7 45	7 26	7 11	6 58	6 46	6 27	6 10	5 54	5 38	5 20	5 00	4 48	4 34
June 3	8 14	7 49	7 29	7 13	7 00	6 48	6 28	6 11	5 54	5 38	5 20	4 59	4 47	4 33
7	8 18	7 52	7 32	7 16	7 02	6 50	6 30	6 12	5 55	5 38	5 20	4 58	4 46	4 31
11	8 22	7 55	7 35	7 18	7 04	6 52	6 31	6 13	5 56	5 39	5 20	4 58	4 45	4 31
15	8 24	7 57	7 37	7 20	7 06	6 54	6 33	6 14	5 57	5 39	5 20	4 58	4 45	4 31
19	8 26	7 59	7 38	7 21	7 07	6 55	6 34	6 15	5 58	5 40	5 21	4 59	4 46	4 31
23	8 27	8 00	7 39	7 22	7 08	6 56	6 34	6 16	5 58	5 41	5 22	5 00	4 47	4 32
27	8 27	8 00	7 40	7 23	7 09	6 56	6 35	6 17	5 59	5 42	5 23	5 01	4 48	4 33
July 1	8 26	8 00	7 39	7 23	7 09	6 57	6 36	6 17	6 00	5 43	5 24	5 02	4 49	4 35
5	8 24	7 58	7 38	7 22	7 08	6 56	6 36	6 18	6 01	5 44	5 25	5 04	4 51	4 37

SUNSET

Lat.	−55°	−50°	−45°	−40°	−35°	−30°	−20°	−10°	0°	+10°	+20°	+30°	+35°	+40°
	h m	h m	h m	h m	h m	h m	h m	h m	h m	h m	h m	h m	h m	h m
Mar. 31	17 46	17 49	17 52	17 54	17 56	17 58	18 02	18 05	18 08	18 11	18 14	18 18	18 20	18 23
Apr. 4	17 36	17 41	17 45	17 48	17 51	17 54	17 58	18 02	18 06	18 10	18 15	18 20	18 23	18 27
8	17 26	17 32	17 37	17 42	17 46	17 49	17 55	18 00	18 05	18 10	18 16	18 23	18 26	18 31
12	17 16	17 24	17 30	17 36	17 40	17 44	17 52	17 58	18 04	18 10	18 17	18 25	18 30	18 35
16	17 07	17 16	17 23	17 30	17 35	17 40	17 48	17 56	18 03	18 11	18 18	18 28	18 33	18 39
20	16 57	17 08	17 17	17 24	17 30	17 36	17 45	17 54	18 02	18 11	18 20	18 30	18 36	18 43
24	16 48	17 00	17 10	17 18	17 25	17 32	17 43	17 52	18 02	18 11	18 21	18 33	18 39	18 47
28	16 39	16 53	17 04	17 13	17 21	17 28	17 40	17 51	18 01	18 11	18 22	18 35	18 43	18 51
May 2	16 31	16 46	16 58	17 08	17 17	17 24	17 38	17 49	18 00	18 12	18 24	18 38	18 46	18 55
6	16 23	16 39	16 53	17 03	17 13	17 21	17 35	17 48	18 00	18 12	18 25	18 40	18 49	18 59
10	16 15	16 33	16 47	16 59	17 09	17 18	17 34	17 47	18 00	18 13	18 27	18 43	18 52	19 03
14	16 08	16 27	16 43	16 55	17 06	17 15	17 32	17 46	18 00	18 14	18 28	18 45	18 56	19 07
18	16 01	16 22	16 38	16 52	17 03	17 13	17 30	17 46	18 00	18 14	18 30	18 48	18 59	19 11
22	15 56	16 18	16 35	16 49	17 01	17 11	17 29	17 45	18 00	18 15	18 32	18 51	19 02	19 14
26	15 50	16 13	16 31	16 46	16 59	17 10	17 28	17 45	18 01	18 16	18 33	18 53	19 04	19 18
30	15 46	16 10	16 29	16 44	16 57	17 08	17 28	17 45	18 01	18 17	18 35	18 55	19 07	19 21
June 3	15 42	16 07	16 27	16 42	16 56	17 08	17 28	17 45	18 02	18 18	18 36	18 57	19 10	19 24
7	15 39	16 05	16 25	16 41	16 55	17 07	17 28	17 46	18 02	18 19	18 38	18 59	19 12	19 26
11	15 37	16 04	16 24	16 41	16 55	17 07	17 28	17 46	18 03	18 21	18 39	19 01	19 14	19 29
15	15 36	16 03	16 24	16 41	16 55	17 07	17 28	17 47	18 04	18 22	18 40	19 02	19 15	19 30
19	15 36	16 04	16 24	16 41	16 55	17 08	17 29	17 47	18 05	18 23	18 42	19 04	19 17	19 32
23	15 37	16 04	16 25	16 42	16 56	17 09	17 30	17 48	18 06	18 23	18 42	19 04	19 18	19 33
27	15 39	16 06	16 27	16 43	16 57	17 10	17 31	17 49	18 07	18 24	18 43	19 05	19 18	19 33
July 1	15 42	16 08	16 28	16 45	16 59	17 11	17 32	17 50	18 07	18 25	18 43	19 05	19 18	19 33
5	15 45	16 11	16 31	16 47	17 01	17 13	17 33	17 51	18 08	18 25	18 44	19 05	19 18	19 32

UNIVERSAL TIME FOR MERIDIAN OF GREENWICH

SUNRISE

Lat.	+40°	+42°	+44°	+46°	+48°	+50°	+52°	+54°	+56°	+58°	+60°	+62°	+64°	+66°
	h m	h m	h m	h m	h m	h m	h m	h m	h m	h m	h m	h m	h m	h m
Mar. 31	5 46	5 45	5 44	5 43	5 41	5 40	5 38	5 36	5 34	5 32	5 30	5 27	5 24	5 20
Apr. 4	5 40	5 39	5 37	5 35	5 33	5 31	5 29	5 27	5 24	5 21	5 18	5 14	5 09	5 04
8	5 34	5 32	5 30	5 28	5 25	5 23	5 20	5 17	5 14	5 10	5 06	5 01	4 55	4 49
12	5 28	5 25	5 23	5 20	5 17	5 14	5 11	5 07	5 03	4 59	4 54	4 48	4 41	4 33
16	5 21	5 19	5 16	5 13	5 10	5 06	5 02	4 58	4 53	4 48	4 42	4 35	4 27	4 18
20	5 16	5 13	5 09	5 06	5 02	4 58	4 54	4 49	4 43	4 37	4 30	4 22	4 13	4 02
24	5 10	5 07	5 03	4 59	4 55	4 50	4 45	4 40	4 33	4 27	4 19	4 10	3 59	3 47
28	5 05	5 01	4 57	4 52	4 48	4 43	4 37	4 31	4 24	4 16	4 07	3 57	3 45	3 31
May 2	4 59	4 55	4 51	4 46	4 41	4 35	4 29	4 22	4 15	4 06	3 56	3 45	3 32	3 15
6	4 55	4 50	4 45	4 40	4 35	4 29	4 22	4 14	4 06	3 57	3 46	3 33	3 18	2 59
10	4 50	4 45	4 40	4 35	4 29	4 22	4 15	4 07	3 58	3 47	3 36	3 22	3 05	2 44
14	4 46	4 41	4 36	4 30	4 23	4 16	4 08	4 00	3 50	3 39	3 26	3 10	2 52	2 28
18	4 43	4 37	4 31	4 25	4 18	4 11	4 02	3 53	3 43	3 31	3 16	3 00	2 39	2 12
22	4 39	4 34	4 28	4 21	4 14	4 06	3 57	3 47	3 36	3 23	3 08	2 50	2 27	1 56
26	4 37	4 31	4 24	4 17	4 10	4 01	3 52	3 42	3 30	3 16	3 00	2 40	2 15	1 39
30	4 34	4 28	4 22	4 14	4 06	3 58	3 48	3 37	3 25	3 10	2 53	2 32	2 04	1 23
June 3	4 33	4 26	4 19	4 12	4 04	3 55	3 45	3 33	3 20	3 05	2 47	2 24	1 54	1 05
7	4 31	4 25	4 18	4 10	4 02	3 53	3 42	3 30	3 17	3 01	2 42	2 18	1 45	0 47
11	4 31	4 24	4 17	4 09	4 00	3 51	3 40	3 28	3 15	2 58	2 39	2 13	1 38	0 24
15	4 31	4 24	4 16	4 09	4 00	3 50	3 39	3 27	3 13	2 57	2 36	2 10	1 33	▭
19	4 31	4 24	4 17	4 09	4 00	3 50	3 39	3 27	3 13	2 56	2 35	2 09	1 31	▭
23	4 32	4 25	4 18	4 10	4 01	3 51	3 40	3 28	3 14	2 57	2 36	2 10	1 31	▭
27	4 33	4 26	4 19	4 11	4 02	3 52	3 42	3 29	3 15	2 59	2 38	2 12	1 35	▭
July 1	4 35	4 28	4 21	4 13	4 04	3 55	3 44	3 32	3 18	3 02	2 42	2 16	1 40	0 14
5	4 37	4 30	4 23	4 15	4 07	3 57	3 47	3 35	3 22	3 06	2 46	2 22	1 48	0 45

SUNSET

Lat.	+40°	+42°	+44°	+46°	+48°	+50°	+52°	+54°	+56°	+58°	+60°	+62°	+64°	+66°
	h m	h m	h m	h m	h m	h m	h m	h m	h m	h m	h m	h m	h m	h m
Mar. 31	18 23	18 24	18 25	18 27	18 28	18 30	18 31	18 33	18 35	18 38	18 40	18 43	18 47	18 51
Apr. 4	18 27	18 28	18 30	18 32	18 34	18 36	18 38	18 41	18 43	18 47	18 50	18 54	18 59	19 04
8	18 31	18 33	18 35	18 37	18 40	18 42	18 45	18 48	18 52	18 56	19 00	19 05	19 11	19 17
12	18 35	18 37	18 40	18 42	18 45	18 48	18 52	18 56	19 00	19 04	19 10	19 16	19 23	19 31
16	18 39	18 42	18 45	18 48	18 51	18 55	18 59	19 03	19 08	19 13	19 20	19 27	19 35	19 44
20	18 43	18 46	18 49	18 53	18 57	19 01	19 06	19 11	19 16	19 22	19 29	19 38	19 47	19 58
24	18 47	18 51	18 54	18 58	19 03	19 07	19 12	19 18	19 24	19 31	19 39	19 49	20 00	20 12
28	18 51	18 55	18 59	19 03	19 08	19 13	19 19	19 25	19 32	19 40	19 49	20 00	20 12	20 27
May 2	18 55	18 59	19 04	19 09	19 14	19 20	19 26	19 33	19 41	19 49	19 59	20 11	20 25	20 42
6	18 59	19 04	19 09	19 14	19 19	19 26	19 32	19 40	19 49	19 58	20 09	20 22	20 38	20 57
10	19 03	19 08	19 13	19 19	19 25	19 32	19 39	19 47	19 56	20 07	20 19	20 33	20 51	21 12
14	19 07	19 12	19 18	19 24	19 30	19 37	19 45	19 54	20 04	20 15	20 29	20 44	21 03	21 28
18	19 11	19 16	19 22	19 28	19 35	19 43	19 51	20 01	20 11	20 24	20 38	20 55	21 16	21 44
22	19 14	19 20	19 26	19 33	19 40	19 48	19 57	20 07	20 19	20 32	20 47	21 06	21 29	22 01
26	19 18	19 24	19 30	19 37	19 45	19 53	20 03	20 13	20 25	20 39	20 55	21 16	21 41	22 18
30	19 21	19 27	19 34	19 41	19 49	19 58	20 08	20 19	20 31	20 46	21 03	21 25	21 53	22 36
June 3	19 24	19 30	19 37	19 45	19 53	20 02	20 12	20 23	20 37	20 52	21 10	21 33	22 04	22 55
7	19 26	19 33	19 40	19 48	19 56	20 05	20 16	20 28	20 41	20 57	21 16	21 41	22 14	23 16
11	19 29	19 35	19 42	19 50	19 59	20 08	20 19	20 31	20 45	21 01	21 21	21 47	22 22	23 43
15	19 30	19 37	19 44	19 52	20 01	20 11	20 22	20 34	20 48	21 05	21 25	21 51	22 28	▭
19	19 32	19 39	19 46	19 54	20 03	20 12	20 23	20 36	20 50	21 07	21 27	21 54	22 32	▭
23	19 33	19 39	19 47	19 55	20 04	20 13	20 24	20 36	20 51	21 08	21 28	21 54	22 33	▭
27	19 33	19 40	19 47	19 55	20 04	20 13	20 24	20 36	20 50	21 07	21 27	21 53	22 30	▭
July 1	19 33	19 39	19 47	19 55	20 03	20 13	20 23	20 35	20 49	21 05	21 25	21 50	22 26	23 43
5	19 32	19 39	19 46	19 53	20 02	20 11	20 22	20 33	20 47	21 03	21 22	21 46	22 19	23 19

▭ indicates Sun continuously above horizon.

UNIVERSAL TIME FOR MERIDIAN OF GREENWICH

SUNRISE

Lat.	−55°	−50°	−45°	−40°	−35°	−30°	−20°	−10°	0°	+10°	+20°	+30°	+35°	+40°
	h m	h m	h m	h m	h m	h m	h m	h m	h m	h m	h m	h m	h m	h m
July 1	8 26	8 00	7 39	7 23	7 09	6 57	6 36	6 17	6 00	5 43	5 24	5 02	4 49	4 35
5	8 24	7 58	7 38	7 22	7 08	6 56	6 36	6 18	6 01	5 44	5 25	5 04	4 51	4 37
9	8 22	7 56	7 37	7 21	7 08	6 56	6 36	6 18	6 02	5 45	5 27	5 06	4 53	4 39
13	8 18	7 54	7 35	7 19	7 06	6 55	6 35	6 18	6 02	5 46	5 28	5 08	4 56	4 42
17	8 14	7 50	7 32	7 17	7 05	6 54	6 35	6 18	6 03	5 47	5 30	5 10	4 58	4 45
21	8 08	7 46	7 29	7 15	7 03	6 52	6 34	6 18	6 03	5 48	5 31	5 12	5 01	4 48
25	8 02	7 42	7 25	7 12	7 00	6 50	6 33	6 17	6 03	5 48	5 33	5 15	5 04	4 52
29	7 56	7 36	7 21	7 08	6 57	6 48	6 31	6 17	6 03	5 49	5 34	5 17	5 07	4 55
Aug. 2	7 49	7 31	7 16	7 04	6 54	6 45	6 29	6 16	6 03	5 50	5 36	5 19	5 10	4 59
6	7 41	7 24	7 11	7 00	6 50	6 42	6 27	6 15	6 02	5 50	5 37	5 22	5 13	5 03
10	7 33	7 18	7 05	6 55	6 46	6 39	6 25	6 13	6 02	5 51	5 38	5 24	5 16	5 06
14	7 25	7 11	7 00	6 50	6 42	6 35	6 23	6 12	6 01	5 51	5 40	5 26	5 19	5 10
18	7 16	7 03	6 53	6 45	6 38	6 31	6 20	6 10	6 01	5 51	5 41	5 29	5 22	5 14
22	7 07	6 56	6 47	6 39	6 33	6 27	6 17	6 08	6 00	5 51	5 42	5 31	5 25	5 18
26	6 57	6 48	6 40	6 33	6 28	6 23	6 14	6 06	5 59	5 51	5 43	5 33	5 28	5 22
30	6 48	6 40	6 33	6 27	6 22	6 18	6 11	6 04	5 57	5 51	5 44	5 36	5 31	5 25
Sept. 3	6 38	6 31	6 26	6 21	6 17	6 13	6 07	6 02	5 56	5 51	5 45	5 38	5 34	5 29
7	6 28	6 23	6 18	6 15	6 12	6 09	6 04	5 59	5 55	5 50	5 46	5 40	5 37	5 33
11	6 18	6 14	6 11	6 08	6 06	6 04	6 00	5 57	5 53	5 50	5 46	5 42	5 40	5 37
15	6 08	6 05	6 03	6 02	6 00	5 59	5 56	5 54	5 52	5 50	5 47	5 44	5 42	5 40
19	5 57	5 57	5 56	5 55	5 54	5 54	5 53	5 52	5 51	5 49	5 48	5 46	5 45	5 44
23	5 47	5 48	5 48	5 48	5 49	5 49	5 49	5 49	5 49	5 49	5 49	5 49	5 48	5 48
27	5 37	5 39	5 41	5 42	5 43	5 44	5 45	5 47	5 48	5 49	5 50	5 51	5 51	5 52
Oct. 1	5 27	5 30	5 33	5 35	5 37	5 39	5 42	5 44	5 47	5 49	5 51	5 53	5 54	5 56
5	5 16	5 21	5 25	5 29	5 32	5 34	5 38	5 42	5 45	5 48	5 52	5 55	5 57	6 00

SUNSET

Lat.	−55°	−50°	−45°	−40°	−35°	−30°	−20°	−10°	0°	+10°	+20°	+30°	+35°	+40°
	h m	h m	h m	h m	h m	h m	h m	h m	h m	h m	h m	h m	h m	h m
July 1	15 42	16 08	16 28	16 45	16 59	17 11	17 32	17 50	18 07	18 25	18 43	19 05	19 18	19 33
5	15 45	16 11	16 31	16 47	17 01	17 13	17 33	17 51	18 08	18 25	18 44	19 05	19 18	19 32
9	15 49	16 14	16 34	16 49	17 03	17 15	17 35	17 52	18 09	18 26	18 43	19 04	19 17	19 31
13	15 54	16 18	16 37	16 52	17 05	17 17	17 36	17 53	18 09	18 26	18 43	19 03	19 15	19 29
17	15 59	16 22	16 40	16 55	17 08	17 19	17 38	17 54	18 10	18 25	18 42	19 02	19 14	19 27
21	16 05	16 27	16 44	16 58	17 10	17 21	17 39	17 55	18 10	18 25	18 41	19 00	19 11	19 24
25	16 11	16 32	16 48	17 02	17 13	17 23	17 40	17 56	18 10	18 25	18 40	18 58	19 09	19 21
29	16 18	16 37	16 53	17 05	17 16	17 25	17 42	17 56	18 10	18 24	18 39	18 56	19 06	19 17
Aug. 2	16 24	16 43	16 57	17 09	17 19	17 28	17 43	17 57	18 10	18 23	18 37	18 53	19 02	19 13
6	16 31	16 48	17 01	17 12	17 22	17 30	17 45	17 57	18 09	18 22	18 35	18 50	18 59	19 09
10	16 38	16 54	17 06	17 16	17 25	17 32	17 46	17 58	18 09	18 20	18 32	18 46	18 54	19 04
14	16 46	16 59	17 11	17 20	17 28	17 35	17 47	17 58	18 08	18 18	18 30	18 43	18 50	18 59
18	16 53	17 05	17 15	17 24	17 31	17 37	17 48	17 58	18 07	18 17	18 27	18 39	18 45	18 53
22	17 00	17 11	17 20	17 27	17 34	17 39	17 49	17 58	18 06	18 15	18 24	18 34	18 41	18 48
26	17 07	17 17	17 25	17 31	17 37	17 42	17 50	17 58	18 05	18 13	18 21	18 30	18 35	18 42
30	17 15	17 23	17 29	17 35	17 39	17 44	17 51	17 58	18 04	18 10	18 17	18 25	18 30	18 35
Sept. 3	17 22	17 29	17 34	17 38	17 42	17 46	17 52	17 57	18 03	18 08	18 14	18 21	18 25	18 29
7	17 29	17 34	17 39	17 42	17 45	17 48	17 53	17 57	18 01	18 06	18 10	18 16	18 19	18 23
11	17 37	17 40	17 43	17 46	17 48	17 50	17 54	17 57	18 00	18 03	18 07	18 11	18 13	18 16
15	17 44	17 46	17 48	17 50	17 51	17 52	17 54	17 57	17 59	18 01	18 03	18 06	18 08	18 10
19	17 51	17 52	17 53	17 53	17 54	17 54	17 55	17 56	17 57	17 58	17 59	18 01	18 02	18 03
23	17 59	17 58	17 58	17 57	17 57	17 57	17 56	17 56	17 56	17 56	17 56	17 56	17 56	17 56
27	18 06	18 04	18 02	18 01	18 00	17 59	17 57	17 56	17 54	17 53	17 52	17 51	17 50	17 50
Oct. 1	18 14	18 10	18 07	18 05	18 03	18 01	17 58	17 55	17 53	17 51	17 49	17 46	17 45	17 43
5	18 22	18 17	18 12	18 09	18 06	18 03	17 59	17 55	17 52	17 48	17 45	17 41	17 39	17 37

UNIVERSAL TIME FOR MERIDIAN OF GREENWICH

SUNRISE

Lat.	+40°	+42°	+44°	+46°	+48°	+50°	+52°	+54°	+56°	+58°	+60°	+62°	+64°	+66°
	h m	h m	h m	h m	h m	h m	h m	h m	h m	h m	h m	h m	h m	h m
July 1	4 35	4 28	4 21	4 13	4 04	3 55	3 44	3 32	3 18	3 02	2 42	2 16	1 40	0 14
5	4 37	4 30	4 23	4 15	4 07	3 57	3 47	3 35	3 22	3 06	2 46	2 22	1 48	0 45
9	4 39	4 33	4 26	4 18	4 10	4 01	3 51	3 39	3 26	3 11	2 52	2 29	1 58	1 05
13	4 42	4 36	4 29	4 22	4 14	4 05	3 55	3 44	3 31	3 17	2 59	2 37	2 08	1 24
17	4 45	4 39	4 32	4 25	4 18	4 09	4 00	3 49	3 37	3 23	3 07	2 46	2 20	1 42
21	4 48	4 42	4 36	4 29	4 22	4 14	4 05	3 55	3 44	3 30	3 15	2 56	2 32	1 59
25	4 52	4 46	4 40	4 34	4 27	4 19	4 11	4 01	3 50	3 38	3 24	3 06	2 45	2 16
29	4 55	4 50	4 44	4 38	4 32	4 24	4 16	4 08	3 57	3 46	3 33	3 17	2 57	2 32
Aug. 2	4 59	4 54	4 49	4 43	4 37	4 30	4 23	4 14	4 05	3 54	3 42	3 27	3 10	2 48
6	5 03	4 58	4 53	4 48	4 42	4 36	4 29	4 21	4 12	4 03	3 51	3 38	3 22	3 03
10	5 06	5 02	4 58	4 53	4 47	4 42	4 35	4 28	4 20	4 11	4 01	3 49	3 35	3 18
14	5 10	5 06	5 02	4 58	4 53	4 47	4 42	4 35	4 28	4 20	4 11	4 00	3 47	3 32
18	5 14	5 10	5 07	5 03	4 58	4 53	4 48	4 42	4 36	4 29	4 20	4 11	4 00	3 46
22	5 18	5 15	5 11	5 08	5 04	4 59	4 55	4 49	4 44	4 37	4 30	4 22	4 12	4 00
26	5 22	5 19	5 16	5 13	5 09	5 05	5 01	4 57	4 52	4 46	4 39	4 32	4 24	4 14
30	5 25	5 23	5 20	5 18	5 15	5 11	5 08	5 04	4 59	4 55	4 49	4 43	4 36	4 27
Sept. 3	5 29	5 27	5 25	5 23	5 20	5 17	5 14	5 11	5 07	5 03	4 58	4 53	4 47	4 40
7	5 33	5 31	5 29	5 27	5 25	5 23	5 21	5 18	5 15	5 12	5 08	5 04	4 59	4 53
11	5 37	5 35	5 34	5 32	5 31	5 29	5 27	5 25	5 23	5 20	5 17	5 14	5 10	5 06
15	5 40	5 40	5 39	5 37	5 36	5 35	5 34	5 32	5 31	5 29	5 27	5 24	5 22	5 19
19	5 44	5 44	5 43	5 42	5 42	5 41	5 40	5 39	5 38	5 37	5 36	5 35	5 33	5 31
23	5 48	5 48	5 48	5 48	5 47	5 47	5 47	5 47	5 46	5 46	5 45	5 45	5 44	5 44
27	5 52	5 52	5 52	5 53	5 53	5 53	5 53	5 54	5 54	5 54	5 55	5 55	5 56	5 56
Oct. 1	5 56	5 56	5 57	5 58	5 58	5 59	6 00	6 01	6 02	6 03	6 04	6 06	6 07	6 09
5	6 00	6 01	6 02	6 03	6 04	6 05	6 07	6 08	6 10	6 12	6 14	6 16	6 19	6 22

SUNSET

Lat.	+40°	+42°	+44°	+46°	+48°	+50°	+52°	+54°	+56°	+58°	+60°	+62°	+64°	+66°
	h m	h m	h m	h m	h m	h m	h m	h m	h m	h m	h m	h m	h m	h m
July 1	19 33	19 39	19 47	19 55	20 03	20 13	20 23	20 35	20 49	21 05	21 25	21 50	22 26	23 43
5	19 32	19 39	19 46	19 53	20 02	20 11	20 22	20 33	20 47	21 03	21 22	21 46	22 19	23 19
9	19 31	19 37	19 44	19 52	20 00	20 09	20 19	20 30	20 43	20 59	21 17	21 40	22 11	23 01
13	19 29	19 35	19 42	19 49	19 57	20 06	20 16	20 27	20 39	20 54	21 11	21 33	22 01	22 43
17	19 27	19 33	19 39	19 46	19 54	20 02	20 12	20 22	20 34	20 48	21 04	21 24	21 50	22 26
21	19 24	19 30	19 36	19 43	19 50	19 58	20 07	20 17	20 28	20 41	20 56	21 15	21 38	22 10
25	19 21	19 26	19 32	19 39	19 46	19 53	20 01	20 11	20 22	20 34	20 48	21 05	21 26	21 54
29	19 17	19 22	19 28	19 34	19 40	19 48	19 56	20 04	20 14	20 26	20 39	20 54	21 13	21 38
Aug. 2	19 13	19 18	19 23	19 29	19 35	19 42	19 49	19 57	20 06	20 17	20 29	20 43	21 00	21 22
6	19 09	19 13	19 18	19 23	19 29	19 35	19 42	19 50	19 58	20 08	20 19	20 32	20 47	21 06
10	19 04	19 08	19 13	19 17	19 23	19 28	19 35	19 42	19 49	19 58	20 08	20 20	20 34	20 50
14	18 59	19 03	19 07	19 11	19 16	19 21	19 27	19 33	19 40	19 48	19 57	20 08	20 20	20 35
18	18 53	18 57	19 00	19 04	19 09	19 13	19 19	19 24	19 31	19 38	19 46	19 55	20 06	20 19
22	18 48	18 51	18 54	18 58	19 01	19 06	19 10	19 15	19 21	19 27	19 34	19 43	19 52	20 03
26	18 42	18 44	18 47	18 50	18 54	18 57	19 02	19 06	19 11	19 16	19 23	19 30	19 38	19 48
30	18 35	18 38	18 40	18 43	18 46	18 49	18 53	18 57	19 01	19 05	19 11	19 17	19 24	19 32
Sept. 3	18 29	18 31	18 33	18 36	18 38	18 41	18 44	18 47	18 50	18 54	18 59	19 04	19 10	19 17
7	18 23	18 24	18 26	18 28	18 30	18 32	18 34	18 37	18 40	18 43	18 47	18 51	18 56	19 01
11	18 16	18 17	18 19	18 20	18 22	18 23	18 25	18 27	18 29	18 32	18 35	18 38	18 41	18 46
15	18 10	18 10	18 11	18 12	18 13	18 15	18 16	18 17	18 19	18 21	18 23	18 25	18 27	18 30
19	18 03	18 03	18 04	18 05	18 05	18 06	18 07	18 07	18 08	18 09	18 10	18 12	18 13	18 15
23	17 56	17 56	17 57	17 57	17 57	17 57	17 57	17 57	17 58	17 58	17 58	17 59	17 59	17 59
27	17 50	17 49	17 49	17 49	17 49	17 48	17 48	17 47	17 47	17 47	17 46	17 45	17 45	17 44
Oct. 1	17 43	17 42	17 42	17 41	17 40	17 39	17 39	17 38	17 36	17 35	17 34	17 32	17 31	17 29
5	17 37	17 36	17 35	17 33	17 32	17 31	17 29	17 28	17 26	17 24	17 22	17 19	17 17	17 14

UNIVERSAL TIME FOR MERIDIAN OF GREENWICH

SUNRISE

Lat.	−55°	−50°	−45°	−40°	−35°	−30°	−20°	−10°	0°	+10°	+20°	+30°	+35°	+40°
	h m	h m	h m	h m	h m	h m	h m	h m	h m	h m	h m	h m	h m	h m
Oct. 1	5 27	5 30	5 33	5 35	5 37	5 39	5 42	5 44	5 47	5 49	5 51	5 53	5 54	5 56
5	5 16	5 21	5 25	5 29	5 32	5 34	5 38	5 42	5 45	5 48	5 52	5 55	5 57	6 00
9	5 06	5 13	5 18	5 22	5 26	5 29	5 35	5 40	5 44	5 48	5 53	5 58	6 01	6 04
13	4 56	5 04	5 11	5 16	5 21	5 25	5 32	5 38	5 43	5 48	5 54	6 00	6 04	6 08
17	4 47	4 56	5 04	5 10	5 16	5 20	5 28	5 36	5 42	5 49	5 55	6 03	6 07	6 12
21	4 37	4 48	4 57	5 04	5 11	5 16	5 26	5 34	5 41	5 49	5 57	6 06	6 11	6 16
25	4 27	4 40	4 50	4 59	5 06	5 12	5 23	5 32	5 41	5 49	5 58	6 08	6 14	6 21
29	4 18	4 33	4 44	4 53	5 01	5 08	5 20	5 31	5 40	5 50	6 00	6 11	6 18	6 25
Nov. 2	4 10	4 25	4 38	4 48	4 57	5 05	5 18	5 30	5 40	5 51	6 02	6 14	6 21	6 30
6	4 01	4 19	4 32	4 44	4 53	5 02	5 16	5 29	5 40	5 52	6 04	6 17	6 25	6 34
10	3 53	4 12	4 27	4 40	4 50	4 59	5 15	5 28	5 40	5 53	6 06	6 21	6 29	6 39
14	3 46	4 07	4 23	4 36	4 47	4 57	5 13	5 28	5 41	5 54	6 08	6 24	6 33	6 43
18	3 39	4 01	4 19	4 33	4 44	4 55	5 12	5 27	5 42	5 55	6 10	6 27	6 37	6 48
22	3 33	3 57	4 15	4 30	4 42	4 53	5 12	5 28	5 42	5 57	6 13	6 30	6 41	6 53
26	3 28	3 53	4 12	4 28	4 41	4 52	5 11	5 28	5 44	5 59	6 15	6 34	6 45	6 57
30	3 23	3 50	4 10	4 26	4 40	4 51	5 11	5 29	5 45	6 01	6 18	6 37	6 48	7 01
Dec. 4	3 20	3 47	4 08	4 25	4 39	4 51	5 12	5 30	5 46	6 03	6 20	6 40	6 52	7 05
8	3 17	3 46	4 07	4 24	4 39	4 51	5 13	5 31	5 48	6 05	6 23	6 43	6 55	7 09
12	3 16	3 45	4 07	4 25	4 39	4 52	5 14	5 33	5 50	6 07	6 25	6 46	6 58	7 12
16	3 15	3 45	4 08	4 25	4 40	4 53	5 15	5 34	5 52	6 09	6 28	6 49	7 01	7 15
20	3 16	3 46	4 09	4 27	4 42	4 55	5 17	5 36	5 54	6 11	6 30	6 51	7 03	7 18
24	3 18	3 48	4 11	4 29	4 44	4 57	5 19	5 38	5 56	6 13	6 32	6 53	7 05	7 20
28	3 21	3 51	4 13	4 31	4 46	4 59	5 21	5 40	5 58	6 15	6 33	6 55	7 07	7 21
32	3 25	3 55	4 17	4 34	4 49	5 02	5 24	5 42	6 00	6 17	6 35	6 56	7 08	7 22
36	3 30	3 59	4 21	4 38	4 52	5 05	5 26	5 44	6 01	6 18	6 36	6 57	7 09	7 22

SUNSET

Lat.	−55°	−50°	−45°	−40°	−35°	−30°	−20°	−10°	0°	+10°	+20°	+30°	+35°	+40°
	h m	h m	h m	h m	h m	h m	h m	h m	h m	h m	h m	h m	h m	h m
Oct. 1	18 14	18 10	18 07	18 05	18 03	18 01	17 58	17 55	17 53	17 51	17 49	17 46	17 45	17 43
5	18 22	18 17	18 12	18 09	18 06	18 03	17 59	17 55	17 52	17 48	17 45	17 41	17 39	17 37
9	18 30	18 23	18 18	18 13	18 09	18 06	18 00	17 55	17 51	17 46	17 42	17 37	17 34	17 30
13	18 38	18 29	18 23	18 17	18 12	18 08	18 01	17 55	17 50	17 44	17 38	17 32	17 28	17 24
17	18 46	18 36	18 28	18 21	18 16	18 11	18 03	17 55	17 49	17 42	17 35	17 28	17 23	17 18
21	18 54	18 42	18 33	18 26	18 19	18 14	18 04	17 56	17 48	17 40	17 32	17 23	17 18	17 13
25	19 02	18 49	18 39	18 30	18 23	18 17	18 06	17 56	17 47	17 39	17 30	17 20	17 14	17 07
29	19 10	18 56	18 44	18 35	18 27	18 20	18 07	17 57	17 47	17 38	17 27	17 16	17 09	17 02
Nov. 2	19 19	19 03	18 50	18 39	18 30	18 23	18 09	17 58	17 47	17 36	17 25	17 13	17 05	16 57
6	19 27	19 10	18 56	18 44	18 34	18 26	18 11	17 59	17 47	17 36	17 23	17 10	17 02	16 53
10	19 36	19 16	19 01	18 49	18 38	18 29	18 13	18 00	17 47	17 35	17 22	17 07	16 58	16 49
14	19 44	19 23	19 07	18 54	18 42	18 33	18 16	18 01	17 48	17 35	17 21	17 05	16 55	16 45
18	19 52	19 30	19 12	18 58	18 46	18 36	18 18	18 03	17 49	17 35	17 20	17 03	16 53	16 42
22	20 00	19 36	19 18	19 03	18 50	18 39	18 21	18 05	17 50	17 35	17 19	17 01	16 51	16 39
26	20 08	19 42	19 23	19 07	18 54	18 43	18 23	18 06	17 51	17 35	17 19	17 00	16 50	16 37
30	20 15	19 48	19 28	19 12	18 58	18 46	18 26	18 08	17 52	17 36	17 19	17 00	16 49	16 36
Dec. 4	20 21	19 53	19 32	19 16	19 01	18 49	18 28	18 10	17 54	17 37	17 20	17 00	16 48	16 35
8	20 27	19 58	19 37	19 19	19 05	18 52	18 31	18 12	17 55	17 39	17 21	17 00	16 48	16 35
12	20 32	20 02	19 40	19 23	19 08	18 55	18 33	18 15	17 57	17 40	17 22	17 01	16 49	16 35
16	20 36	20 06	19 44	19 26	19 11	18 58	18 36	18 17	17 59	17 42	17 23	17 02	16 50	16 36
20	20 39	20 09	19 46	19 28	19 13	19 00	18 38	18 19	18 01	17 44	17 25	17 04	16 52	16 37
24	20 41	20 11	19 48	19 30	19 15	19 02	18 40	18 21	18 03	17 46	17 27	17 06	16 54	16 39
28	20 41	20 12	19 49	19 31	19 16	19 03	18 41	18 23	18 05	17 48	17 29	17 08	16 56	16 42
32	20 41	20 12	19 50	19 32	19 17	19 05	18 43	18 24	18 07	17 50	17 32	17 11	16 59	16 45
36	20 39	20 11	19 49	19 32	19 18	19 05	18 44	18 26	18 09	17 52	17 34	17 14	17 02	16 48

UNIVERSAL TIME FOR MERIDIAN OF GREENWICH

SUNRISE

Lat.	+40°	+42°	+44°	+46°	+48°	+50°	+52°	+54°	+56°	+58°	+60°	+62°	+64°	+66°
	h m	h m	h m	h m	h m	h m	h m	h m	h m	h m	h m	h m	h m	h m
Oct. 1	5 56	5 56	5 57	5 58	5 58	5 59	6 00	6 01	6 02	6 03	6 04	6 06	6 07	6 09
5	6 00	6 01	6 02	6 03	6 04	6 05	6 07	6 08	6 10	6 12	6 14	6 16	6 19	6 22
9	6 04	6 05	6 07	6 08	6 10	6 12	6 14	6 16	6 18	6 21	6 24	6 27	6 31	6 35
13	6 08	6 10	6 11	6 13	6 16	6 18	6 20	6 23	6 26	6 30	6 33	6 38	6 42	6 48
17	6 12	6 14	6 16	6 19	6 21	6 24	6 27	6 31	6 34	6 39	6 43	6 48	6 54	7 01
21	6 16	6 19	6 21	6 24	6 27	6 31	6 34	6 38	6 43	6 48	6 53	6 59	7 07	7 15
25	6 21	6 24	6 27	6 30	6 33	6 37	6 41	6 46	6 51	6 57	7 03	7 10	7 19	7 29
29	6 25	6 28	6 32	6 36	6 40	6 44	6 49	6 54	7 00	7 06	7 13	7 22	7 31	7 43
Nov. 2	6 30	6 33	6 37	6 41	6 46	6 51	6 56	7 02	7 08	7 15	7 24	7 33	7 44	7 57
6	6 34	6 38	6 42	6 47	6 52	6 57	7 03	7 09	7 17	7 25	7 34	7 44	7 57	8 12
10	6 39	6 43	6 48	6 53	6 58	7 04	7 10	7 17	7 25	7 34	7 44	7 56	8 10	8 27
14	6 43	6 48	6 53	6 58	7 04	7 10	7 17	7 25	7 34	7 43	7 54	8 07	8 23	8 42
18	6 48	6 53	6 58	7 04	7 10	7 17	7 24	7 33	7 42	7 52	8 04	8 19	8 36	8 57
22	6 53	6 58	7 03	7 09	7 16	7 23	7 31	7 40	7 50	8 01	8 14	8 29	8 48	9 12
26	6 57	7 02	7 08	7 15	7 22	7 29	7 37	7 47	7 57	8 09	8 23	8 40	9 01	9 27
30	7 01	7 07	7 13	7 20	7 27	7 35	7 44	7 53	8 04	8 17	8 32	8 50	9 12	9 42
Dec. 4	7 05	7 11	7 17	7 24	7 32	7 40	7 49	7 59	8 11	8 24	8 40	8 59	9 23	9 56
8	7 09	7 15	7 21	7 29	7 36	7 45	7 54	8 05	8 17	8 31	8 47	9 07	9 33	10 09
12	7 12	7 18	7 25	7 32	7 40	7 49	7 59	8 09	8 22	8 36	8 53	9 14	9 41	10 20
16	7 15	7 21	7 28	7 36	7 44	7 52	8 02	8 13	8 26	8 40	8 58	9 19	9 47	10 28
20	7 18	7 24	7 31	7 38	7 46	7 55	8 05	8 16	8 29	8 44	9 01	9 23	9 51	10 34
24	7 20	7 26	7 33	7 40	7 48	7 57	8 07	8 18	8 31	8 46	9 03	9 25	9 53	10 36
28	7 21	7 27	7 34	7 41	7 49	7 58	8 08	8 19	8 32	8 46	9 04	9 25	9 53	10 34
32	7 22	7 28	7 35	7 42	7 50	7 59	8 08	8 19	8 31	8 46	9 03	9 23	9 50	10 29
36	7 22	7 28	7 35	7 42	7 50	7 58	8 07	8 18	8 30	8 44	9 00	9 20	9 45	10 21

SUNSET

Lat.	+40°	+42°	+44°	+46°	+48°	+50°	+52°	+54°	+56°	+58°	+60°	+62°	+64°	+66°
	h m	h m	h m	h m	h m	h m	h m	h m	h m	h m	h m	h m	h m	h m
Oct. 1	17 43	17 42	17 42	17 41	17 40	17 39	17 39	17 38	17 36	17 35	17 34	17 32	17 31	17 29
5	17 37	17 36	17 35	17 33	17 32	17 31	17 29	17 28	17 26	17 24	17 22	17 19	17 17	17 14
9	17 30	17 29	17 27	17 26	17 24	17 22	17 20	17 18	17 16	17 13	17 10	17 07	17 03	16 58
13	17 24	17 22	17 21	17 18	17 16	17 14	17 11	17 09	17 05	17 02	16 58	16 54	16 49	16 43
17	17 18	17 16	17 14	17 11	17 09	17 06	17 03	16 59	16 55	16 51	16 47	16 41	16 35	16 28
21	17 13	17 10	17 07	17 04	17 01	16 58	16 54	16 50	16 46	16 41	16 35	16 29	16 22	16 13
25	17 07	17 04	17 01	16 58	16 54	16 50	16 46	16 41	16 36	16 30	16 24	16 17	16 08	15 58
29	17 02	16 59	16 55	16 51	16 47	16 43	16 38	16 33	16 27	16 20	16 13	16 05	15 55	15 43
Nov. 2	16 57	16 53	16 49	16 45	16 41	16 36	16 31	16 25	16 18	16 11	16 03	15 53	15 42	15 29
6	16 53	16 49	16 44	16 40	16 35	16 29	16 23	16 17	16 10	16 02	15 52	15 42	15 29	15 14
10	16 49	16 44	16 40	16 35	16 29	16 23	16 17	16 10	16 02	15 53	15 43	15 31	15 17	15 00
14	16 45	16 40	16 35	16 30	16 24	16 18	16 11	16 03	15 55	15 45	15 34	15 21	15 05	14 46
18	16 42	16 37	16 32	16 26	16 20	16 13	16 05	15 57	15 48	15 37	15 25	15 11	14 54	14 32
22	16 39	16 34	16 28	16 22	16 16	16 08	16 01	15 52	15 42	15 30	15 17	15 02	14 43	14 19
26	16 37	16 32	16 26	16 19	16 12	16 05	15 56	15 47	15 37	15 25	15 10	14 54	14 33	14 06
30	16 36	16 30	16 24	16 17	16 10	16 02	15 53	15 43	15 32	15 19	15 05	14 47	14 24	13 55
Dec. 4	16 35	16 29	16 22	16 16	16 08	16 00	15 51	15 40	15 29	15 15	15 00	14 41	14 17	13 44
8	16 35	16 28	16 22	16 15	16 07	15 58	15 49	15 38	15 26	15 13	14 56	14 36	14 11	13 34
12	16 35	16 29	16 22	16 15	16 07	15 58	15 48	15 38	15 25	15 11	14 54	14 33	14 06	13 27
16	16 36	16 29	16 23	16 15	16 07	15 58	15 49	15 38	15 25	15 10	14 53	14 32	14 04	13 22
20	16 37	16 31	16 24	16 17	16 09	16 00	15 50	15 39	15 26	15 11	14 54	14 32	14 04	13 21
24	16 39	16 33	16 26	16 19	16 11	16 02	15 52	15 41	15 28	15 13	14 56	14 34	14 06	13 23
28	16 42	16 36	16 29	16 21	16 13	16 05	15 55	15 44	15 31	15 17	14 59	14 38	14 10	13 29
32	16 45	16 39	16 32	16 25	16 17	16 08	15 59	15 48	15 35	15 21	15 04	14 44	14 17	13 38
36	16 48	16 42	16 36	16 29	16 21	16 13	16 03	15 53	15 41	15 27	15 10	14 51	14 25	13 50

UNIVERSAL TIME FOR MERIDIAN OF GREENWICH

MOONRISE

Lat.	−55°	−50°	−45°	−40°	−35°	−30°	−20°	−10°	0°	+10°	+20°	+30°	+35°	+40°
	h m	h m	h m	h m	h m	h m	h m	h m	h m	h m	h m	h m	h m	h m
Jan. 0	18 20	17 44	17 18	16 57	16 39	16 25	15 59	15 38	15 18	14 58	14 37	14 12	13 58	13 42
1	19 44	19 02	18 31	18 08	17 48	17 32	17 04	16 40	16 18	15 55	15 32	15 04	14 48	14 29
2	20 48	20 05	19 34	19 10	18 51	18 34	18 05	17 41	17 18	16 55	16 31	16 03	15 46	15 26
3	21 29	20 52	20 24	20 02	19 44	19 28	19 01	18 38	18 17	17 55	17 32	17 05	16 49	16 31
4	21 54	21 24	21 01	20 43	20 27	20 14	19 51	19 31	19 12	18 53	18 33	18 10	17 56	17 40
5	22 09	21 47	21 30	21 15	21 03	20 52	20 34	20 18	20 03	19 47	19 31	19 12	19 01	18 48
6	22 20	22 04	21 52	21 42	21 33	21 25	21 12	21 00	20 49	20 37	20 25	20 12	20 04	19 54
7	22 27	22 18	22 10	22 04	21 59	21 54	21 45	21 38	21 31	21 24	21 17	21 08	21 03	20 58
8	22 33	22 30	22 27	22 24	22 22	22 20	22 17	22 14	22 11	22 09	22 06	22 02	22 00	21 58
9	22 39	22 41	22 42	22 43	22 44	22 45	22 47	22 49	22 50	22 52	22 53	22 55	22 56	22 58
10	22 44	22 52	22 58	23 03	23 07	23 11	23 18	23 24	23 29	23 35	23 41	23 48	23 52	23 57
11	22 51	23 04	23 15	23 24	23 31	23 38	23 50	.. ..	.. ..	.. ..	.. ..	.. ..	.. ..	.. ..
12	23 00	23 19	23 35	23 47	23 58	.. ..	.. ..	0 00	0 10	0 19	0 30	0 42	0 49	0 57
13	23 12	23 39	23 59	.. ..	.. ..	0 08	0 24	0 39	0 52	1 06	1 21	1 38	1 48	2 00
14	23 31	.. ..	.. ..	0 16	0 30	0 42	1 03	1 21	1 39	1 56	2 15	2 37	2 49	3 04
15	.. ..	0 05	0 30	0 51	1 08	1 22	1 47	2 09	2 29	2 50	3 12	3 37	3 52	4 10
16	0 02	0 43	1 12	1 35	1 54	2 10	2 38	3 02	3 24	3 46	4 11	4 39	4 55	5 15
17	0 52	1 35	2 06	2 30	2 50	3 07	3 35	3 59	4 22	4 45	5 09	5 38	5 55	6 15
18	2 05	2 45	3 13	3 36	3 54	4 10	4 37	5 00	5 22	5 43	6 06	6 33	6 48	7 06
19	3 34	4 06	4 30	4 49	5 05	5 18	5 41	6 01	6 20	6 38	6 58	7 21	7 34	7 49
20	5 10	5 33	5 51	6 05	6 17	6 28	6 46	7 01	7 16	7 30	7 46	8 03	8 13	8 25
21	6 47	7 01	7 12	7 21	7 29	7 36	7 48	7 59	8 08	8 18	8 28	8 40	8 47	8 54
22	8 20	8 27	8 32	8 36	8 40	8 43	8 49	8 54	8 58	9 03	9 08	9 14	9 17	9 20
23	9 52	9 51	9 51	9 50	9 49	9 49	9 48	9 48	9 47	9 46	9 46	9 45	9 45	9 45
24	11 24	11 15	11 09	11 03	10 59	10 54	10 47	10 41	10 35	10 30	10 24	10 17	10 13	10 09

MOONSET

Lat.	−55°	−50°	−45°	−40°	−35°	−30°	−20°	−10°	0°	+10°	+20°	+30°	+35°	+40°
	h m	h m	h m	h m	h m	h m	h m	h m	h m	h m	h m	h m	h m	h m
Jan. 0	0 11	0 40	1 02	1 20	1 35	1 48	2 10	2 29	2 48	3 06	3 26	3 49	4 02	4 18
1	0 36	1 13	1 41	2 02	2 20	2 36	3 02	3 25	3 46	4 07	4 30	4 57	5 13	5 31
2	1 17	2 00	2 31	2 54	3 14	3 31	3 59	4 24	4 47	5 09	5 34	6 02	6 19	6 39
3	2 19	3 02	3 32	3 56	4 15	4 32	5 00	5 24	5 47	6 09	6 33	7 00	7 17	7 36
4	3 38	4 15	4 42	5 03	5 21	5 36	6 02	6 24	6 44	7 04	7 26	7 51	8 05	8 22
5	5 05	5 34	5 55	6 13	6 27	6 40	7 01	7 20	7 37	7 54	8 12	8 33	8 45	8 59
6	6 31	6 51	7 07	7 20	7 31	7 41	7 58	8 12	8 25	8 38	8 52	9 08	9 17	9 28
7	7 52	8 06	8 16	8 25	8 33	8 39	8 50	9 00	9 09	9 18	9 28	9 39	9 45	9 52
8	9 10	9 17	9 22	9 27	9 31	9 35	9 41	9 46	9 51	9 55	10 01	10 06	10 10	10 13
9	10 25	10 26	10 26	10 27	10 28	10 28	10 29	10 29	10 30	10 31	10 31	10 32	10 32	10 33
10	11 39	11 34	11 30	11 26	11 23	11 21	11 16	11 13	11 09	11 05	11 02	10 57	10 55	10 52
11	12 53	12 42	12 33	12 26	12 20	12 14	12 05	11 56	11 48	11 41	11 33	11 23	11 18	11 12
12	14 10	13 52	13 38	13 27	13 17	13 09	12 54	12 42	12 30	12 18	12 06	11 51	11 43	11 34
13	15 30	15 05	14 46	14 30	14 17	14 06	13 47	13 30	13 14	12 59	12 42	12 23	12 12	12 00
14	16 52	16 19	15 55	15 36	15 19	15 06	14 42	14 21	14 02	13 44	13 23	13 00	12 46	12 31
15	18 12	17 32	17 04	16 41	16 23	16 07	15 40	15 17	14 55	14 34	14 11	13 44	13 28	13 10
16	19 21	18 38	18 08	17 44	17 24	17 07	16 39	16 14	15 52	15 29	15 04	14 36	14 19	13 59
17	20 12	19 32	19 03	18 40	18 21	18 04	17 37	17 13	16 51	16 28	16 04	15 36	15 19	15 00
18	20 45	20 12	19 47	19 27	19 10	18 56	18 31	18 10	17 50	17 29	17 08	16 42	16 27	16 10
19	21 06	20 41	20 22	20 06	19 53	19 41	19 21	19 03	18 47	18 30	18 12	17 51	17 39	17 25
20	21 19	21 03	20 49	20 38	20 29	20 21	20 06	19 53	19 41	19 29	19 16	19 01	18 52	18 42
21	21 29	21 20	21 12	21 06	21 00	20 56	20 47	20 40	20 32	20 25	20 17	20 08	20 03	19 57
22	21 37	21 35	21 33	21 31	21 29	21 28	21 26	21 24	21 22	21 20	21 17	21 15	21 13	21 12
23	21 45	21 49	21 52	21 55	21 57	22 00	22 03	22 07	22 10	22 13	22 16	22 20	22 22	22 25
24	21 53	22 04	22 13	22 20	22 26	22 32	22 42	22 51	22 59	23 07	23 16	23 26	23 32	23 38

.. .. indicates phenomenon will occur the next day.

UNIVERSAL TIME FOR MERIDIAN OF GREENWICH

MOONRISE

Lat.	+40°	+42°	+44°	+46°	+48°	+50°	+52°	+54°	+56°	+58°	+60°	+62°	+64°	+66°
	h m	h m	h m	h m	h m	h m	h m	h m	h m	h m	h m	h m	h m	h m
Jan. 0	13 42	13 34	13 26	13 18	13 08	12 58	12 46	12 33	12 17	11 59	11 37	11 09	10 28	□
1	14 29	14 20	14 11	14 01	13 50	13 37	13 23	13 07	12 47	12 23	11 52	11 00	□	□
2	15 26	15 17	15 07	14 57	14 45	14 32	14 16	13 59	13 37	13 10	12 32	□	□	□
3	16 31	16 23	16 13	16 03	15 52	15 40	15 26	15 09	14 50	14 26	13 53	12 58	□	□
4	17 40	17 32	17 25	17 16	17 07	16 56	16 45	16 31	16 16	15 58	15 35	15 05	14 19	□
5	18 48	18 43	18 36	18 30	18 23	18 15	18 06	17 56	17 45	17 32	17 16	16 58	16 35	16 04
6	19 54	19 50	19 46	19 41	19 36	19 31	19 24	19 18	19 10	19 01	18 51	18 40	18 26	18 10
7	20 58	20 55	20 52	20 50	20 46	20 43	20 39	20 35	20 31	20 26	20 20	20 14	20 06	19 57
8	21 58	21 57	21 56	21 55	21 54	21 53	21 51	21 50	21 48	21 46	21 44	21 42	21 39	21 36
9	22 58	22 58	22 59	22 59	23 00	23 01	23 02	23 02	23 03	23 05	23 06	23 07	23 09	23 11
10	23 57	23 59	.. ..	.. ..	.. ..	.. ..	.. ..	.. ..	.. ..	.. ..	.. ..	.. ..	.. ..	.. ..
11	.. ..	.. ..	0 01	0 03	0 06	0 09	0 12	0 15	0 19	0 23	0 28	0 33	0 39	0 46
12	0 57	1 01	1 05	1 09	1 13	1 18	1 23	1 29	1 36	1 43	1 52	2 02	2 13	2 27
13	2 00	2 05	2 10	2 16	2 23	2 30	2 38	2 46	2 56	3 08	3 21	3 37	3 56	4 21
14	3 04	3 11	3 18	3 26	3 34	3 44	3 54	4 06	4 20	4 36	4 56	5 21	5 55	6 57
15	4 10	4 18	4 27	4 36	4 47	4 58	5 12	5 27	5 45	6 07	6 36	7 18	■	■
16	5 15	5 24	5 33	5 44	5 56	6 09	6 24	6 42	7 03	7 30	8 10	■	■	■
17	6 15	6 24	6 33	6 44	6 56	7 09	7 25	7 42	8 04	8 32	9 11	■	■	■
18	7 06	7 15	7 24	7 33	7 44	7 56	8 09	8 25	8 43	9 05	9 34	10 16	■	■
19	7 49	7 56	8 04	8 12	8 20	8 30	8 41	8 53	9 06	9 22	9 41	10 05	10 36	11 24
20	8 25	8 30	8 35	8 41	8 48	8 55	9 02	9 11	9 20	9 31	9 44	9 58	10 15	10 36
21	8 54	8 58	9 01	9 05	9 09	9 14	9 19	9 24	9 30	9 36	9 44	9 52	10 02	10 13
22	9 20	9 22	9 24	9 25	9 27	9 29	9 32	9 34	9 37	9 40	9 43	9 47	9 51	9 56
23	9 45	9 44	9 44	9 44	9 44	9 44	9 43	9 43	9 43	9 43	9 42	9 42	9 42	9 41
24	10 09	10 07	10 05	10 03	10 01	9 58	9 55	9 52	9 49	9 46	9 42	9 37	9 32	9 26

MOONSET

Lat.	+40°	+42°	+44°	+46°	+48°	+50°	+52°	+54°	+56°	+58°	+60°	+62°	+64°	+66°
	h m	h m	h m	h m	h m	h m	h m	h m	h m	h m	h m	h m	h m	h m
Jan. 0	4 18	4 25	4 32	4 40	4 49	4 59	5 11	5 23	5 38	5 56	6 17	6 45	7 25	□
1	5 31	5 40	5 49	5 58	6 09	6 22	6 36	6 52	7 11	7 35	8 06	8 57	□	□
2	6 39	6 47	6 57	7 08	7 20	7 33	7 48	8 06	8 27	8 54	9 33	□	□	□
3	7 36	7 44	7 54	8 04	8 15	8 28	8 42	8 59	9 19	9 43	10 16	11 11	□	□
4	8 22	8 30	8 38	8 47	8 56	9 07	9 19	9 33	9 49	10 08	10 31	11 01	11 48	□
5	8 59	9 05	9 11	9 18	9 26	9 35	9 44	9 55	10 06	10 20	10 36	10 55	11 19	11 51
6	9 28	9 32	9 37	9 43	9 48	9 55	10 01	10 09	10 17	10 27	10 38	10 50	11 05	11 23
7	9 52	9 55	9 58	10 02	10 06	10 10	10 14	10 19	10 25	10 31	10 38	10 45	10 54	11 05
8	10 13	10 15	10 17	10 18	10 20	10 23	10 25	10 27	10 30	10 33	10 37	10 41	10 45	10 50
9	10 33	10 33	10 33	10 33	10 34	10 34	10 34	10 34	10 35	10 35	10 36	10 36	10 36	10 37
10	10 52	10 51	10 49	10 48	10 47	10 45	10 43	10 41	10 39	10 37	10 34	10 31	10 28	10 24
11	11 12	11 09	11 06	11 03	11 00	10 57	10 53	10 48	10 44	10 39	10 33	10 26	10 19	10 10
12	11 34	11 30	11 25	11 21	11 16	11 10	11 04	10 57	10 50	10 42	10 32	10 21	10 09	9 53
13	12 00	11 54	11 48	11 41	11 34	11 27	11 18	11 09	10 58	10 46	10 32	10 16	9 55	9 30
14	12 31	12 24	12 16	12 08	11 59	11 49	11 38	11 25	11 11	10 54	10 34	10 09	9 34	8 31
15	13 10	13 02	12 53	12 43	12 32	12 20	12 07	11 51	11 33	11 10	10 41	9 58	■	■
16	13 59	13 50	13 41	13 30	13 18	13 05	12 49	12 32	12 10	11 43	11 03	■	■	■
17	15 00	14 51	14 41	14 31	14 19	14 06	13 50	13 33	13 11	12 44	12 05	■	■	■
18	16 10	16 02	15 53	15 44	15 33	15 22	15 08	14 53	14 36	14 14	13 46	13 04	■	■
19	17 25	17 18	17 12	17 04	16 56	16 47	16 37	16 25	16 12	15 57	15 39	15 16	14 46	13 58
20	18 42	18 37	18 32	18 27	18 21	18 15	18 08	18 00	17 52	17 42	17 30	17 17	17 01	16 41
21	19 57	19 55	19 52	19 49	19 45	19 42	19 38	19 34	19 29	19 23	19 17	19 10	19 02	18 53
22	21 12	21 11	21 10	21 09	21 08	21 07	21 06	21 05	21 03	21 02	21 00	20 58	20 56	20 53
23	22 25	22 26	22 27	22 28	22 30	22 31	22 33	22 35	22 37	22 39	22 41	22 44	22 47	22 51
24	23 38	23 41	23 44	23 48	23 52	23 56	.. ..	.. ..	.. ..	.. ..	.. ..	.. ..	.. ..	.. ..

□ indicates Moon continuously above horizon.
■ indicates Moon continuously below horizon.
.. .. indicates phenomenon will occur the next day.

UNIVERSAL TIME FOR MERIDIAN OF GREENWICH

MOONRISE

Lat.	−55°	−50°	−45°	−40°	−35°	−30°	−20°	−10°	0°	+10°	+20°	+30°	+35°	+40°
	h m	h m	h m	h m	h m	h m	h m	h m	h m	h m	h m	h m	h m	h m
Jan. 23	9 52	9 51	9 51	9 50	9 49	9 49	9 48	9 48	9 47	9 46	9 46	9 45	9 45	9 45
24	11 24	11 15	11 09	11 03	10 59	10 54	10 47	10 41	10 35	10 30	10 24	10 17	10 13	10 09
25	12 56	12 40	12 28	12 17	12 08	12 00	11 47	11 36	11 25	11 14	11 03	10 51	10 43	10 35
26	14 31	14 06	13 47	13 32	13 19	13 08	12 49	12 32	12 17	12 02	11 46	11 27	11 17	11 05
27	16 05	15 31	15 06	14 46	14 30	14 16	13 52	13 31	13 12	12 53	12 33	12 09	11 56	11 40
28	17 32	16 50	16 21	15 58	15 39	15 22	14 55	14 31	14 10	13 48	13 25	12 58	12 42	12 24
29	18 41	17 57	17 26	17 02	16 42	16 25	15 56	15 31	15 09	14 46	14 21	13 53	13 36	13 16
30	19 29	18 48	18 19	17 56	17 37	17 21	16 53	16 29	16 07	15 45	15 21	14 53	14 37	14 17
31	19 58	19 25	19 00	18 40	18 23	18 09	17 44	17 23	17 03	16 43	16 21	15 56	15 41	15 24
Feb. 1	20 16	19 50	19 31	19 15	19 01	18 49	18 29	18 11	17 54	17 38	17 20	16 59	16 47	16 32
2	20 27	20 09	19 55	19 43	19 33	19 24	19 08	18 55	18 42	18 29	18 15	17 59	17 50	17 39
3	20 36	20 24	20 15	20 07	20 00	19 54	19 44	19 34	19 26	19 17	19 08	18 57	18 51	18 44
4	20 42	20 36	20 32	20 28	20 24	20 21	20 16	20 11	20 07	20 02	19 58	19 52	19 49	19 46
5	20 48	20 47	20 47	20 47	20 47	20 47	20 47	20 46	20 46	20 46	20 46	20 46	20 46	20 46
6	20 53	20 58	21 03	21 06	21 09	21 12	21 17	21 21	21 25	21 29	21 34	21 39	21 42	21 45
7	20 59	21 10	21 19	21 26	21 33	21 38	21 48	21 57	22 05	22 13	22 22	22 32	22 38	22 45
8	21 07	21 24	21 37	21 48	21 58	22 06	22 21	22 34	22 46	22 59	23 12	23 27	23 36	23 46
9	21 17	21 41	21 59	22 14	22 27	22 38	22 58	23 15	23 31	23 47	.. ..	.. ..	.. ..	.. ..
10	21 32	22 03	22 27	22 45	23 01	23 15	23 39	23 59	.. ..	.. ..	0 04	0 24	0 36	0 49
11	21 56	22 35	23 02	23 24	23 43	23 59	.. ..	.. ..	0 18	0 38	0 58	1 23	1 37	1 54
12	22 36	23 19	23 50	.. ..	.. ..	.. ..	0 25	0 49	1 10	1 32	1 55	2 23	2 39	2 58
13	23 37	.. ..	.. ..	0 13	0 33	0 50	1 18	1 43	2 06	2 29	2 53	3 22	3 39	3 59
14	.. ..	0 20	0 50	1 13	1 33	1 49	2 17	2 42	3 04	3 26	3 50	4 18	4 35	4 54
15	0 59	1 35	2 02	2 23	2 40	2 55	3 21	3 42	4 03	4 23	4 45	5 09	5 24	5 41
16	2 33	3 01	3 22	3 38	3 52	4 05	4 25	4 43	5 00	5 17	5 34	5 55	6 06	6 20

MOONSET

Lat.	−55°	−50°	−45°	−40°	−35°	−30°	−20°	−10°	0°	+10°	+20°	+30°	+35°	+40°
	h m	h m	h m	h m	h m	h m	h m	h m	h m	h m	h m	h m	h m	h m
Jan. 23	21 45	21 49	21 52	21 55	21 57	22 00	22 03	22 07	22 10	22 13	22 16	22 20	22 22	22 25
24	21 53	22 04	22 13	22 20	22 26	22 32	22 42	22 51	22 59	23 07	23 16	23 26	23 32	23 38
25	22 03	22 21	22 35	22 48	22 58	23 07	23 23	23 36	23 49	.. ..	.. ..	.. ..	.. ..	.. ..
26	22 16	22 43	23 03	23 20	23 34	23 46	.. ..	.. ..	.. ..	0 02	0 16	0 32	0 42	0 53
27	22 37	23 12	23 38	23 59	.. ..	.. ..	0 07	0 25	0 43	1 00	1 19	1 40	1 53	2 07
28	23 11	23 53	.. ..	.. ..	0 16	0 31	0 57	1 19	1 39	2 00	2 22	2 48	3 03	3 21
29	.. ..	.. ..	0 23	0 47	1 06	1 23	1 51	2 15	2 38	3 00	3 25	3 53	4 09	4 29
30	0 05	0 49	1 20	1 44	2 04	2 21	2 50	3 14	3 37	4 00	4 24	4 52	5 09	5 28
31	1 18	1 58	2 26	2 49	3 07	3 23	3 50	4 13	4 34	4 55	5 18	5 44	6 00	6 18
Feb. 1	2 42	3 14	3 38	3 57	4 13	4 26	4 50	5 09	5 28	5 46	6 06	6 29	6 42	6 57
2	4 08	4 32	4 50	5 05	5 17	5 28	5 47	6 03	6 18	6 32	6 48	7 06	7 16	7 28
3	5 31	5 48	6 00	6 11	6 20	6 28	6 41	6 52	7 03	7 14	7 25	7 38	7 46	7 54
4	6 51	7 00	7 08	7 14	7 19	7 24	7 32	7 39	7 46	7 52	7 59	8 07	8 11	8 16
5	8 07	8 10	8 13	8 15	8 17	8 18	8 21	8 24	8 26	8 28	8 31	8 33	8 35	8 36
6	9 21	9 19	9 16	9 15	9 13	9 12	9 09	9 07	9 05	9 03	9 01	8 59	8 57	8 56
7	10 36	10 27	10 20	10 14	10 09	10 05	9 57	9 50	9 44	9 38	9 32	9 24	9 20	9 15
8	11 51	11 36	11 24	11 14	11 06	10 59	10 46	10 35	10 25	10 14	10 03	9 51	9 44	9 36
9	13 09	12 47	12 30	12 16	12 04	11 54	11 37	11 21	11 07	10 53	10 38	10 21	10 11	10 00
10	14 30	14 00	13 38	13 20	13 05	12 52	12 30	12 11	11 53	11 35	11 16	10 55	10 42	10 28
11	15 50	15 13	14 46	14 24	14 07	13 51	13 26	13 03	12 43	12 22	12 00	11 35	11 20	11 02
12	17 04	16 21	15 51	15 27	15 08	14 51	14 23	13 59	13 37	13 14	12 50	12 22	12 05	11 46
13	18 03	17 20	16 50	16 26	16 06	15 49	15 21	14 56	14 34	14 11	13 46	13 17	13 00	12 40
14	18 43	18 06	17 38	17 17	16 59	16 43	16 17	15 54	15 32	15 10	14 47	14 20	14 04	13 45
15	19 08	18 40	18 17	18 00	17 44	17 31	17 09	16 49	16 30	16 12	15 51	15 28	15 14	14 58
16	19 25	19 04	18 48	18 35	18 24	18 13	17 56	17 41	17 26	17 12	16 56	16 38	16 27	16 15

.. .. indicates phenomenon will occur the next day.

UNIVERSAL TIME FOR MERIDIAN OF GREENWICH

MOONRISE

Lat.	+40°	+42°	+44°	+46°	+48°	+50°	+52°	+54°	+56°	+58°	+60°	+62°	+64°	+66°
	h m	h m	h m	h m	h m	h m	h m	h m	h m	h m	h m	h m	h m	h m
Jan. 23	9 45	9 44	9 44	9 44	9 44	9 44	9 43	9 43	9 43	9 43	9 42	9 42	9 42	9 41
24	10 09	10 07	10 05	10 03	10 01	9 58	9 55	9 52	9 49	9 46	9 42	9 37	9 32	9 26
25	10 35	10 31	10 27	10 23	10 19	10 14	10 09	10 03	9 56	9 49	9 41	9 32	9 21	9 09
26	11 05	10 59	10 54	10 47	10 41	10 33	10 25	10 16	10 06	9 55	9 42	9 27	9 08	8 45
27	11 40	11 33	11 26	11 17	11 09	10 59	10 48	10 35	10 21	10 05	9 45	9 20	8 47	7 52
28	12 24	12 15	12 06	11 56	11 46	11 34	11 20	11 04	10 46	10 24	9 55	9 12	▭	▭
29	13 16	13 07	12 58	12 47	12 35	12 22	12 07	11 49	11 28	11 00	10 22	▭	▭	▭
30	14 17	14 09	13 59	13 49	13 37	13 24	13 09	12 52	12 31	12 05	11 29	▭	▭	▭
31	15 24	15 16	15 08	14 59	14 49	14 37	14 25	14 10	13 53	13 32	13 06	12 28	▭	▭
Feb. 1	16 32	16 26	16 19	16 12	16 04	15 55	15 45	15 33	15 21	15 06	14 48	14 25	13 56	13 11
2	17 39	17 35	17 30	17 24	17 18	17 12	17 04	16 56	16 47	16 37	16 25	16 11	15 54	15 33
3	18 44	18 41	18 37	18 34	18 30	18 26	18 21	18 16	18 10	18 04	17 57	17 48	17 38	17 27
4	19 46	19 44	19 43	19 41	19 39	19 37	19 34	19 32	19 29	19 26	19 23	19 19	19 14	19 09
5	20 46	20 46	20 46	20 46	20 46	20 46	20 46	20 46	20 46	20 46	20 46	20 45	20 45	20 45
6	21 45	21 47	21 48	21 50	21 52	21 54	21 56	21 58	22 01	22 04	22 07	22 11	22 16	22 21
7	22 45	22 48	22 51	22 55	22 59	23 03	23 07	23 12	23 17	23 24	23 31	23 39	23 48	23 59
8	23 46	23 51	23 56	.. ..	.. ..	.. ..	.. ..	.. ..	.. ..	.. ..	.. ..	.. ..	.. ..	.. ..
9	.. ..	.. ..	.. ..	0 01	0 07	0 13	0 20	0 27	0 36	0 46	0 57	1 10	1 26	1 46
10	0 49	0 55	1 02	1 09	1 17	1 25	1 35	1 45	1 58	2 12	2 29	2 49	3 16	3 55
11	1 54	2 01	2 09	2 18	2 28	2 39	2 51	3 05	3 21	3 41	4 06	4 39	5 41	▬
12	2 58	3 06	3 16	3 26	3 37	3 50	4 05	4 22	4 42	5 08	5 43	6 58	▬	▬
13	3 59	4 08	4 18	4 29	4 41	4 55	5 10	5 29	5 51	6 20	7 02	▬	▬	▬
14	4 54	5 03	5 12	5 23	5 34	5 47	6 02	6 19	6 39	7 05	7 39	8 46	▬	▬
15	5 41	5 49	5 57	6 06	6 16	6 27	6 39	6 53	7 09	7 28	7 51	8 22	9 11	▬
16	6 20	6 26	6 32	6 39	6 47	6 55	7 04	7 15	7 26	7 39	7 55	8 13	8 36	9 06

MOONSET

Lat.	+40°	+42°	+44°	+46°	+48°	+50°	+52°	+54°	+56°	+58°	+60°	+62°	+64°	+66°
	h m	h m	h m	h m	h m	h m	h m	h m	h m	h m	h m	h m	h m	h m
Jan. 23	22 25	22 26	22 27	22 28	22 30	22 31	22 33	22 35	22 37	22 39	22 41	22 44	22 47	22 51
24	23 38	23 41	23 44	23 48	23 52	23 56	.. ..	.. ..	.. ..	.. ..	.. ..	.. ..	.. ..	.. ..
25	.. ..	.. ..	.. ..	.. ..	.. ..	.. ..	0 00	0 05	0 10	0 16	0 23	0 31	0 40	0 51
26	0 53	0 57	1 03	1 08	1 14	1 21	1 28	1 36	1 46	1 56	2 08	2 22	2 40	3 01
27	2 07	2 14	2 21	2 29	2 37	2 47	2 57	3 09	3 22	3 38	3 57	4 21	4 53	5 48
28	3 21	3 29	3 37	3 47	3 57	4 09	4 22	4 38	4 56	5 18	5 47	6 29	▭	▭
29	4 29	4 38	4 47	4 58	5 10	5 23	5 38	5 56	6 17	6 44	7 22	▭	▭	▭
30	5 28	5 37	5 47	5 58	6 09	6 22	6 37	6 55	7 16	7 42	8 19	▭	▭	▭
31	6 18	6 26	6 34	6 44	6 54	7 06	7 19	7 34	7 52	8 13	8 40	9 18	▭	▭
Feb. 1	6 57	7 04	7 11	7 19	7 27	7 37	7 47	7 59	8 12	8 28	8 47	9 10	9 40	10 26
2	7 28	7 33	7 39	7 45	7 52	7 59	8 07	8 15	8 25	8 36	8 49	9 04	9 22	9 44
3	7 54	7 58	8 02	8 06	8 11	8 16	8 21	8 27	8 33	8 41	8 49	8 59	9 10	9 23
4	8 16	8 19	8 21	8 23	8 26	8 29	8 32	8 36	8 39	8 44	8 48	8 54	9 00	9 07
5	8 36	8 37	8 38	8 39	8 40	8 41	8 42	8 43	8 44	8 46	8 47	8 49	8 51	8 53
6	8 56	8 55	8 54	8 53	8 53	8 52	8 51	8 50	8 48	8 47	8 46	8 44	8 42	8 40
7	9 15	9 13	9 11	9 08	9 06	9 03	9 00	8 56	8 53	8 49	8 44	8 39	8 33	8 26
8	9 36	9 32	9 29	9 24	9 20	9 15	9 10	9 04	8 58	8 51	8 43	8 34	8 23	8 11
9	10 00	9 55	9 49	9 43	9 37	9 30	9 23	9 14	9 05	8 54	8 42	8 28	8 11	7 50
10	10 28	10 21	10 14	10 07	9 58	9 49	9 39	9 28	9 15	9 00	8 43	8 22	7 54	7 15
11	11 02	10 55	10 46	10 37	10 27	10 16	10 03	9 49	9 32	9 12	8 47	8 12	7 10	▬
12	11 46	11 37	11 28	11 17	11 06	10 53	10 38	10 21	10 00	9 34	8 59	7 43	▬	▬
13	12 40	12 31	12 21	12 10	11 58	11 45	11 29	11 11	10 49	10 20	9 37	▬	▬	▬
14	13 45	13 37	13 27	13 17	13 06	12 53	12 39	12 22	12 02	11 37	11 02	9 56	▬	▬
15	14 58	14 51	14 43	14 35	14 25	14 15	14 03	13 50	13 34	13 16	12 53	12 22	11 34	▬
16	16 15	16 10	16 04	15 57	15 50	15 43	15 34	15 25	15 14	15 02	14 47	14 30	14 08	13 39

▭ indicates Moon continuously above horizon.
▬ indicates Moon continuously below horizon.
.. .. indicates phenomenon will occur the next day.

UNIVERSAL TIME FOR MERIDIAN OF GREENWICH

MOONRISE

Lat.	−55°	−50°	−45°	−40°	−35°	−30°	−20°	−10°	0°	+10°	+20°	+30°	+35°	+40°
	h m	h m	h m	h m	h m	h m	h m	h m	h m	h m	h m	h m	h m	h m
Feb. 15	0 59	1 35	2 02	2 23	2 40	2 55	3 21	3 42	4 03	4 23	4 45	5 09	5 24	5 41
16	2 33	3 01	3 22	3 38	3 52	4 05	4 25	4 43	5 00	5 17	5 34	5 55	6 06	6 20
17	4 12	4 30	4 44	4 56	5 06	5 15	5 30	5 43	5 55	6 07	6 20	6 34	6 43	6 52
18	5 49	5 59	6 07	6 14	6 19	6 24	6 33	6 40	6 47	6 54	7 02	7 10	7 15	7 20
19	7 25	7 27	7 29	7 30	7 32	7 33	7 35	7 36	7 38	7 40	7 42	7 44	7 45	7 46
20	9 00	8 55	8 50	8 47	8 43	8 41	8 36	8 32	8 28	8 25	8 21	8 16	8 14	8 11
21	10 36	10 22	10 12	10 03	9 56	9 49	9 38	9 28	9 19	9 10	9 01	8 50	8 44	8 37
22	12 13	11 51	11 34	11 20	11 08	10 58	10 41	10 26	10 12	9 58	9 43	9 27	9 17	9 06
23	13 50	13 18	12 55	12 36	12 21	12 07	11 45	11 25	11 07	10 49	10 30	10 08	9 55	9 40
24	15 21	14 41	14 12	13 50	13 31	13 15	12 49	12 26	12 05	11 43	11 21	10 55	10 39	10 22
25	16 37	15 52	15 21	14 57	14 37	14 20	13 51	13 26	13 03	12 41	12 16	11 48	11 31	11 12
26	17 30	16 47	16 17	15 53	15 34	15 17	14 49	14 24	14 02	13 39	13 15	12 46	12 30	12 10
27	18 03	17 27	17 01	16 40	16 22	16 07	15 41	15 19	14 58	14 37	14 14	13 48	13 33	13 15
28	18 23	17 55	17 34	17 16	17 02	16 49	16 27	16 08	15 50	15 32	15 13	14 50	14 37	14 22
Mar. 1	18 36	18 16	18 00	17 46	17 35	17 25	17 07	16 52	16 38	16 24	16 08	15 51	15 40	15 28
2	18 45	18 31	18 20	18 11	18 03	17 56	17 44	17 33	17 23	17 12	17 01	16 49	16 41	16 33
3	18 52	18 44	18 38	18 32	18 28	18 24	18 16	18 10	18 04	17 58	17 52	17 45	17 40	17 35
4	18 58	18 55	18 54	18 52	18 51	18 50	18 48	18 46	18 44	18 42	18 41	18 39	18 37	18 36
5	19 03	19 06	19 09	19 11	19 13	19 15	19 18	19 21	19 23	19 26	19 28	19 32	19 33	19 36
6	19 09	19 18	19 25	19 31	19 36	19 41	19 49	19 56	20 02	20 09	20 16	20 25	20 30	20 35
7	19 16	19 30	19 42	19 52	20 00	20 08	20 21	20 32	20 43	20 54	21 06	21 19	21 27	21 36
8	19 25	19 46	20 02	20 16	20 28	20 38	20 56	21 11	21 26	21 41	21 56	22 15	22 25	22 38
9	19 37	20 06	20 27	20 45	20 59	21 12	21 34	21 54	22 12	22 30	22 49	23 12	23 26	23 41
10	19 57	20 33	20 59	21 20	21 37	21 52	22 18	22 40	23 01	23 22	23 45	.. ..	.. ..	.. ..
11	20 29	21 10	21 40	22 03	22 23	22 39	23 07	23 32	23 54	.. ..	.. ..	0 11	0 26	0 45

MOONSET

Lat.	−55°	−50°	−45°	−40°	−35°	−30°	−20°	−10°	0°	+10°	+20°	+30°	+35°	+40°
	h m	h m	h m	h m	h m	h m	h m	h m	h m	h m	h m	h m	h m	h m
Feb. 15	19 08	18 40	18 17	18 00	17 44	17 31	17 09	16 49	16 30	16 12	15 51	15 28	15 14	14 58
16	19 25	19 04	18 48	18 35	18 24	18 13	17 56	17 41	17 26	17 12	16 56	16 38	16 27	16 15
17	19 36	19 24	19 14	19 05	18 58	18 51	18 40	18 30	18 20	18 10	18 00	17 48	17 41	17 33
18	19 45	19 40	19 35	19 32	19 28	19 25	19 20	19 16	19 12	19 07	19 02	18 57	18 54	18 50
19	19 53	19 55	19 56	19 57	19 58	19 58	20 00	20 01	20 02	20 03	20 04	20 05	20 06	20 07
20	20 01	20 09	20 16	20 22	20 27	20 31	20 39	20 46	20 52	20 59	21 05	21 13	21 18	21 23
21	20 10	20 26	20 39	20 49	20 58	21 06	21 20	21 32	21 44	21 55	22 08	22 22	22 30	22 40
22	20 23	20 47	21 05	21 21	21 34	21 45	22 04	22 22	22 38	22 54	23 11	23 31	23 43	23 56
23	20 41	21 14	21 38	21 58	22 15	22 29	22 53	23 14	23 34	23 54	.. ..	.. ..	.. ..	.. ..
24	21 11	21 51	22 21	22 44	23 03	23 19	23 47	.. ..	.. ..	.. ..	0 15	0 40	0 55	1 12
25	21 58	22 42	23 14	23 38	23 58	.. ..	.. ..	0 10	0 33	0 55	1 19	1 47	2 03	2 22
26	23 05	23 47	.. ..	.. ..	.. ..	0 15	0 44	1 09	1 32	1 55	2 19	2 48	3 05	3 25
27	.. ..	.. ..	0 17	0 40	0 59	1 16	1 43	2 07	2 29	2 51	3 14	3 42	3 58	4 16
28	0 26	1 01	1 26	1 47	2 03	2 18	2 43	3 04	3 23	3 43	4 04	4 28	4 42	4 58
Mar. 1	1 51	2 17	2 38	2 54	3 08	3 20	3 40	3 57	4 14	4 30	4 47	5 06	5 18	5 31
2	3 14	3 33	3 48	4 00	4 10	4 19	4 34	4 48	5 00	5 12	5 25	5 40	5 48	5 58
3	4 34	4 46	4 56	5 03	5 10	5 16	5 26	5 35	5 43	5 51	5 59	6 09	6 15	6 21
4	5 51	5 57	6 01	6 05	6 08	6 11	6 16	6 20	6 24	6 27	6 31	6 36	6 38	6 41
5	7 06	7 06	7 05	7 05	7 05	7 04	7 04	7 03	7 03	7 02	7 02	7 01	7 01	7 01
6	8 20	8 14	8 09	8 04	8 01	7 57	7 52	7 47	7 42	7 37	7 32	7 27	7 24	7 20
7	9 36	9 23	9 13	9 04	8 57	8 51	8 40	8 31	8 22	8 13	8 04	7 53	7 47	7 40
8	10 53	10 33	10 18	10 06	9 55	9 46	9 30	9 16	9 03	8 51	8 37	8 22	8 13	8 02
9	12 12	11 45	11 25	11 08	10 54	10 42	10 22	10 04	9 48	9 31	9 14	8 54	8 42	8 29
10	13 32	12 57	12 32	12 12	11 55	11 40	11 16	10 55	10 35	10 16	9 55	9 30	9 16	9 00
11	14 48	14 07	13 37	13 14	12 55	12 39	12 12	11 48	11 26	11 04	10 41	10 14	9 58	9 39

.. .. indicates phenomenon will occur the next day.

UNIVERSAL TIME FOR MERIDIAN OF GREENWICH

MOONRISE

Lat.	+40°	+42°	+44°	+46°	+48°	+50°	+52°	+54°	+56°	+58°	+60°	+62°	+64°	+66°
	h m	h m	h m	h m	h m	h m	h m	h m	h m	h m	h m	h m	h m	h m
Feb. 15	5 41	5 49	5 57	6 06	6 16	6 27	6 39	6 53	7 09	7 28	7 51	8 22	9 11	■
16	6 20	6 26	6 32	6 39	6 47	6 55	7 04	7 15	7 26	7 39	7 55	8 13	8 36	9 06
17	6 52	6 57	7 01	7 06	7 11	7 17	7 23	7 30	7 37	7 46	7 56	8 07	8 20	8 35
18	7 20	7 23	7 25	7 28	7 31	7 34	7 38	7 41	7 45	7 50	7 55	8 01	8 08	8 15
19	7 46	7 47	7 47	7 48	7 49	7 49	7 50	7 51	7 52	7 53	7 54	7 56	7 57	7 59
20	8 11	8 10	8 08	8 07	8 06	8 04	8 02	8 00	7 58	7 56	7 53	7 50	7 47	7 43
21	8 37	8 34	8 31	8 27	8 24	8 20	8 15	8 10	8 05	7 59	7 53	7 45	7 37	7 27
22	9 06	9 01	8 56	8 51	8 45	8 38	8 31	8 23	8 14	8 04	7 53	7 40	7 24	7 05
23	9 40	9 34	9 27	9 19	9 11	9 02	8 51	8 40	8 27	8 12	7 55	7 33	7 06	6 27
24	10 22	10 14	10 05	9 55	9 45	9 34	9 21	9 06	8 48	8 28	8 01	7 25	6 07	□
25	11 12	11 03	10 53	10 42	10 31	10 18	10 03	9 45	9 24	8 58	8 21	□	□	□
26	12 10	12 01	11 51	11 41	11 29	11 16	11 00	10 42	10 21	9 53	9 14	□	□	□
27	13 15	13 06	12 58	12 48	12 37	12 25	12 12	11 56	11 37	11 14	10 44	9 58	□	□
28	14 22	14 15	14 07	13 59	13 51	13 41	13 30	13 17	13 03	12 46	12 26	11 59	11 22	□
Mar. 1	15 28	15 23	15 17	15 11	15 05	14 57	14 49	14 40	14 29	14 18	14 04	13 47	13 27	13 01
2	16 33	16 29	16 25	16 21	16 17	16 11	16 06	16 00	15 53	15 45	15 36	15 26	15 14	15 00
3	17 35	17 33	17 31	17 29	17 26	17 23	17 20	17 17	17 13	17 08	17 04	16 58	16 52	16 44
4	18 36	18 35	18 35	18 34	18 33	18 33	18 32	18 31	18 30	18 29	18 27	18 26	18 24	18 22
5	19 36	19 37	19 38	19 39	19 40	19 41	19 42	19 44	19 45	19 47	19 49	19 52	19 54	19 58
6	20 35	20 38	20 40	20 43	20 46	20 49	20 53	20 57	21 01	21 06	21 12	21 18	21 26	21 35
7	21 36	21 40	21 44	21 49	21 54	21 59	22 05	22 12	22 19	22 28	22 37	22 48	23 02	23 18
8	22 38	22 43	22 49	22 56	23 03	23 10	23 19	23 28	23 39	23 52	.. ..	.. ..	.. ..	.. ..
9	23 41	23 48	23 56	.. ..	.. ..	.. ..	.. ..	.. ..	.. ..	.. ..	0 06	0 24	0 46	1 15
10	.. ..	.. ..	.. ..	0 04	0 13	0 23	0 34	0 47	1 02	1 19	1 40	2 08	2 49	■
11	0 45	0 53	1 02	1 12	1 22	1 34	1 48	2 04	2 23	2 46	3 17	4 07	■	■

MOONSET

Lat.	+40°	+42°	+44°	+46°	+48°	+50°	+52°	+54°	+56°	+58°	+60°	+62°	+64°	+66°
	h m	h m	h m	h m	h m	h m	h m	h m	h m	h m	h m	h m	h m	h m
Feb. 15	14 58	14 51	14 43	14 35	14 25	14 15	14 03	13 50	13 34	13 16	12 53	12 22	11 34	■
16	16 15	16 10	16 04	15 57	15 50	15 43	15 34	15 25	15 14	15 02	14 47	14 30	14 08	13 39
17	17 33	17 30	17 26	17 22	17 17	17 12	17 07	17 01	16 55	16 47	16 39	16 29	16 17	16 03
18	18 50	18 49	18 47	18 45	18 43	18 41	18 39	18 36	18 33	18 30	18 26	18 22	18 18	18 12
19	20 07	20 07	20 07	20 08	20 08	20 09	20 09	20 10	20 10	20 11	20 12	20 13	20 14	20 15
20	21 23	21 25	21 28	21 30	21 33	21 36	21 39	21 43	21 47	21 52	21 57	22 03	22 10	22 18
21	22 40	22 44	22 48	22 53	22 58	23 04	23 11	23 18	23 26	23 34	23 45	23 57	.. ..	.. ..
22	23 56	.. ..	.. ..	.. ..	.. ..	.. ..	.. ..	.. ..	.. ..	.. ..	.. ..	.. ..	0 11	0 28
23	.. ..	0 02	0 09	0 16	0 24	0 32	0 42	0 53	1 05	1 19	1 36	1 56	2 23	3 01
24	1 12	1 19	1 28	1 37	1 47	1 58	2 11	2 25	2 42	3 02	3 28	4 04	5 21	□
25	2 22	2 31	2 41	2 51	3 03	3 16	3 31	3 48	4 09	4 35	5 12	□	□	□
26	3 25	3 34	3 43	3 54	4 06	4 19	4 35	4 53	5 14	5 42	6 22	□	□	□
27	4 16	4 25	4 34	4 44	4 55	5 07	5 21	5 37	5 56	6 19	6 49	7 36	□	□
28	4 58	5 05	5 13	5 21	5 30	5 41	5 52	6 05	6 20	6 37	6 59	7 26	8 04	□
Mar. 1	5 31	5 37	5 43	5 49	5 57	6 05	6 13	6 23	6 34	6 47	7 01	7 19	7 40	8 07
2	5 58	6 02	6 07	6 12	6 17	6 23	6 29	6 36	6 43	6 52	7 02	7 13	7 26	7 42
3	6 21	6 24	6 27	6 30	6 33	6 37	6 41	6 45	6 50	6 55	7 01	7 08	7 16	7 25
4	6 41	6 43	6 44	6 45	6 47	6 49	6 51	6 53	6 55	6 57	7 00	7 03	7 06	7 10
5	7 01	7 01	7 00	7 00	7 00	7 00	7 00	6 59	6 59	6 59	6 58	6 58	6 57	6 57
6	7 20	7 18	7 17	7 15	7 13	7 11	7 08	7 06	7 03	7 00	6 57	6 53	6 49	6 43
7	7 40	7 37	7 34	7 30	7 27	7 23	7 18	7 13	7 08	7 02	6 55	6 48	6 39	6 29
8	8 02	7 58	7 53	7 48	7 42	7 36	7 30	7 22	7 14	7 05	6 54	6 42	6 28	6 11
9	8 29	8 23	8 16	8 09	8 02	7 53	7 44	7 34	7 23	7 10	6 54	6 36	6 13	5 43
10	9 00	8 53	8 45	8 36	8 27	8 16	8 05	7 51	7 36	7 18	6 56	6 28	5 47	■
11	9 39	9 31	9 21	9 11	9 00	8 48	8 34	8 18	7 58	7 35	7 03	6 13	■	■

□ indicates Moon continuously above horizon.
■ indicates Moon continuously below horizon.
.. .. indicates phenomenon will occur the next day.

UNIVERSAL TIME FOR MERIDIAN OF GREENWICH

MOONRISE

Lat.	−55°	−50°	−45°	−40°	−35°	−30°	−20°	−10°	0°	+10°	+20°	+30°	+35°	+40°
	h m	h m	h m	h m	h m	h m	h m	h m	h m	h m	h m	h m	h m	h m
Mar. 9	19 37	20 06	20 27	20 45	20 59	21 12	21 34	21 54	22 12	22 30	22 49	23 12	23 26	23 41
10	19 57	20 33	20 59	21 20	21 37	21 52	22 18	22 40	23 01	23 22	23 45	.. ..	.. ..	.. ..
11	20 29	21 10	21 40	22 03	22 23	22 39	23 07	23 32	23 54	.. ..	.. ..	0 11	0 26	0 45
12	21 18	22 02	22 33	22 57	23 17	23 34	.. ..	.. ..	.. ..	0 17	0 41	1 09	1 26	1 46
13	22 30	23 09	23 38	.. ..	.. ..	.. ..	0 02	0 27	0 50	1 13	1 37	2 06	2 23	2 42
14	23 57	.. ..	.. ..	0 01	0 19	0 35	1 02	1 25	1 47	2 08	2 31	2 58	3 13	3 32
15	.. ..	0 29	0 53	1 12	1 28	1 41	2 04	2 25	2 43	3 02	3 22	3 44	3 58	4 13
16	1 32	1 55	2 13	2 27	2 39	2 50	3 08	3 24	3 38	3 53	4 08	4 26	4 36	4 48
17	3 09	3 23	3 35	3 44	3 52	3 59	4 11	4 22	4 31	4 41	4 51	5 03	5 10	5 18
18	4 46	4 52	4 57	5 02	5 05	5 08	5 14	5 19	5 23	5 28	5 32	5 38	5 41	5 44
19	6 23	6 21	6 20	6 19	6 18	6 18	6 16	6 15	6 14	6 13	6 12	6 11	6 11	6 10
20	8 01	7 51	7 44	7 38	7 32	7 28	7 19	7 12	7 06	7 00	6 53	6 45	6 41	6 36
21	9 41	9 23	9 09	8 57	8 47	8 39	8 24	8 11	8 00	7 48	7 36	7 22	7 14	7 05
22	11 22	10 55	10 34	10 17	10 03	9 51	9 30	9 12	8 56	8 40	8 22	8 02	7 51	7 38
23	13 00	12 23	11 56	11 35	11 18	11 03	10 37	10 15	9 55	9 35	9 13	8 49	8 34	8 17
24	14 25	13 42	13 11	12 47	12 27	12 11	11 42	11 18	10 56	10 33	10 09	9 41	9 25	9 06
25	15 28	14 44	14 13	13 49	13 29	13 12	12 43	12 19	11 56	11 33	11 08	10 40	10 23	10 03
26	16 07	15 29	15 01	14 39	14 21	14 05	13 38	13 15	12 53	12 32	12 08	11 41	11 25	11 07
27	16 31	16 00	15 37	15 19	15 03	14 49	14 26	14 06	13 47	13 28	13 08	12 44	12 30	12 14
28	16 45	16 23	16 05	15 50	15 38	15 27	15 08	14 52	14 36	14 21	14 04	13 45	13 34	13 21
29	16 55	16 39	16 27	16 16	16 07	15 59	15 45	15 33	15 21	15 10	14 57	14 43	14 35	14 26
30	17 02	16 53	16 45	16 38	16 32	16 27	16 19	16 11	16 03	15 56	15 48	15 39	15 34	15 28
31	17 08	17 04	17 01	16 58	16 56	16 54	16 50	16 47	16 43	16 40	16 37	16 33	16 31	16 29
Apr. 1	17 14	17 15	17 16	17 17	17 18	17 19	17 20	17 21	17 22	17 24	17 25	17 26	17 27	17 28
2	17 19	17 26	17 32	17 37	17 41	17 44	17 51	17 56	18 02	18 07	18 13	18 19	18 23	18 27

MOONSET

Lat.	−55°	−50°	−45°	−40°	−35°	−30°	−20°	−10°	0°	+10°	+20°	+30°	+35°	+40°
	h m	h m	h m	h m	h m	h m	h m	h m	h m	h m	h m	h m	h m	h m
Mar. 9	12 12	11 45	11 25	11 08	10 54	10 42	10 22	10 04	9 48	9 31	9 14	8 54	8 42	8 29
10	13 32	12 57	12 32	12 12	11 55	11 40	11 16	10 55	10 35	10 16	9 55	9 30	9 16	9 00
11	14 48	14 07	13 37	13 14	12 55	12 39	12 12	11 48	11 26	11 04	10 41	10 14	9 58	9 39
12	15 52	15 08	14 37	14 13	13 54	13 37	13 08	12 43	12 21	11 58	11 33	11 04	10 47	10 27
13	16 39	15 59	15 29	15 06	14 47	14 31	14 03	13 39	13 17	12 55	12 30	12 02	11 46	11 26
14	17 10	16 36	16 12	15 52	15 35	15 20	14 56	14 34	14 14	13 54	13 32	13 06	12 51	12 34
15	17 29	17 04	16 45	16 29	16 16	16 04	15 44	15 26	15 10	14 53	14 35	14 14	14 01	13 47
16	17 43	17 26	17 13	17 01	16 52	16 43	16 29	16 16	16 04	15 51	15 38	15 23	15 14	15 04
17	17 52	17 43	17 36	17 29	17 24	17 19	17 11	17 03	16 56	16 49	16 41	16 32	16 27	16 21
18	18 01	17 59	17 57	17 55	17 54	17 53	17 51	17 49	17 47	17 45	17 43	17 41	17 40	17 38
19	18 09	18 14	18 18	18 21	18 24	18 26	18 31	18 35	18 38	18 42	18 46	18 51	18 53	18 56
20	18 18	18 30	18 40	18 48	18 55	19 01	19 12	19 22	19 31	19 40	19 50	20 01	20 08	20 15
21	18 29	18 49	19 05	19 18	19 30	19 40	19 57	20 12	20 26	20 40	20 55	21 13	21 23	21 35
22	18 45	19 14	19 36	19 54	20 10	20 23	20 46	21 05	21 24	21 42	22 02	22 25	22 39	22 55
23	19 11	19 49	20 16	20 38	20 57	21 12	21 39	22 02	22 24	22 45	23 09	23 36	23 52	.. ..
24	19 52	20 36	21 07	21 31	21 51	22 08	22 37	23 02	23 24	23 47	.. ..	.. ..	.. ..	0 10
25	20 54	21 38	22 09	22 32	22 52	23 09	23 37	.. ..	.. ..	.. ..	0 12	0 41	0 58	1 17
26	22 13	22 50	23 17	23 39	23 56	.. ..	.. ..	0 01	0 24	0 46	1 10	1 38	1 54	2 14
27	23 37	.. ..	.. ..	.. ..	.. ..	0 11	0 37	0 59	1 20	1 40	2 02	2 27	2 41	2 58
28	.. ..	0 06	0 28	0 46	1 01	1 13	1 35	1 54	2 11	2 28	2 47	3 08	3 20	3 34
29	1 01	1 22	1 39	1 52	2 03	2 13	2 30	2 45	2 58	3 12	3 26	3 42	3 52	4 02
30	2 22	2 36	2 47	2 56	3 04	3 11	3 22	3 32	3 42	3 51	4 01	4 13	4 19	4 26
31	3 39	3 46	3 52	3 57	4 02	4 05	4 12	4 18	4 23	4 28	4 34	4 40	4 43	4 47
Apr. 1	4 54	4 55	4 56	4 57	4 58	4 59	5 00	5 01	5 02	5 03	5 04	5 05	5 06	5 07
2	6 08	6 03	6 00	5 57	5 54	5 52	5 48	5 44	5 41	5 38	5 35	5 31	5 28	5 26

.. .. indicates phenomenon will occur the next day.

UNIVERSAL TIME FOR MERIDIAN OF GREENWICH

MOONRISE

Lat.	+40°	+42°	+44°	+46°	+48°	+50°	+52°	+54°	+56°	+58°	+60°	+62°	+64°	+66°
	h m	h m	h m	h m	h m	h m	h m	h m	h m	h m	h m	h m	h m	h m
Mar. 9	23 41	23 48	23 56	.. ..	.. ..	.. ..	.. ..	.. ..	.. ..	.. ..	0 06	0 24	0 46	1 15
10	.. ..	.. ..	.. ..	0 04	0 13	0 23	0 34	0 47	1 02	1 19	1 40	2 08	2 49	■
11	0 45	0 53	1 02	1 12	1 22	1 34	1 48	2 04	2 23	2 46	3 17	4 07	■	■
12	1 46	1 55	2 05	2 16	2 27	2 41	2 56	3 14	3 36	4 05	4 46	■	■	■
13	2 42	2 51	3 01	3 12	3 24	3 37	3 53	4 11	4 32	5 00	5 40	■	■	■
14	3 32	3 40	3 49	3 59	4 09	4 21	4 35	4 50	5 09	5 31	6 00	6 42	■	■
15	4 13	4 20	4 27	4 35	4 44	4 54	5 05	5 17	5 30	5 46	6 06	6 29	7 01	7 51
16	4 48	4 53	4 59	5 05	5 11	5 18	5 26	5 34	5 44	5 55	6 07	6 22	6 39	7 01
17	5 18	5 21	5 25	5 28	5 33	5 37	5 42	5 47	5 53	6 00	6 07	6 16	6 26	6 37
18	5 44	5 46	5 48	5 49	5 51	5 53	5 55	5 58	6 00	6 03	6 07	6 10	6 14	6 19
19	6 10	6 10	6 09	6 09	6 09	6 08	6 08	6 07	6 07	6 06	6 06	6 05	6 04	6 03
20	6 36	6 34	6 32	6 29	6 26	6 24	6 21	6 17	6 14	6 09	6 05	6 00	5 54	5 47
21	7 05	7 00	6 56	6 52	6 47	6 41	6 35	6 29	6 22	6 14	6 05	5 54	5 42	5 28
22	7 38	7 32	7 25	7 19	7 11	7 03	6 54	6 45	6 33	6 21	6 06	5 49	5 27	5 00
23	8 17	8 10	8 02	7 53	7 43	7 33	7 21	7 07	6 52	6 33	6 11	5 41	4 58	□
24	9 06	8 57	8 48	8 37	8 26	8 13	7 59	7 42	7 22	6 57	6 24	5 27	□	□
25	10 03	9 54	9 44	9 33	9 21	9 08	8 53	8 35	8 13	7 45	7 05	□	□	□
26	11 07	10 58	10 49	10 39	10 28	10 15	10 01	9 45	9 25	9 00	8 27	7 28	□	□
27	12 14	12 07	11 59	11 50	11 41	11 30	11 18	11 05	10 49	10 31	10 07	9 37	8 48	□
28	13 21	13 15	13 09	13 02	12 54	12 46	12 37	12 27	12 16	12 02	11 47	11 28	11 04	10 31
29	14 26	14 21	14 17	14 12	14 07	14 01	13 55	13 47	13 40	13 31	13 20	13 08	12 54	12 37
30	15 28	15 25	15 23	15 20	15 16	15 13	15 09	15 04	15 00	14 54	14 48	14 41	14 33	14 24
31	16 29	16 27	16 26	16 25	16 24	16 22	16 21	16 19	16 17	16 15	16 12	16 10	16 06	16 03
Apr. 1	17 28	17 28	17 29	17 29	17 30	17 30	17 31	17 32	17 32	17 33	17 34	17 35	17 36	17 38
2	18 27	18 29	18 31	18 34	18 36	18 39	18 41	18 45	18 48	18 52	18 56	19 01	19 07	19 14

MOONSET

Lat.	+40°	+42°	+44°	+46°	+48°	+50°	+52°	+54°	+56°	+58°	+60°	+62°	+64°	+66°
	h m	h m	h m	h m	h m	h m	h m	h m	h m	h m	h m	h m	h m	h m
Mar. 9	8 29	8 23	8 16	8 09	8 02	7 53	7 44	7 34	7 23	7 10	6 54	6 36	6 13	5 43
10	9 00	8 53	8 45	8 36	8 27	8 16	8 05	7 51	7 36	7 18	6 56	6 28	5 47	■
11	9 39	9 31	9 21	9 11	9 00	8 48	8 34	8 18	7 58	7 35	7 03	6 13	■	■
12	10 27	10 18	10 09	9 58	9 46	9 32	9 17	8 58	8 36	8 08	7 27	■	■	■
13	11 26	11 17	11 07	10 57	10 45	10 32	10 16	9 59	9 37	9 09	8 29	■	■	■
14	12 34	12 26	12 17	12 08	11 57	11 45	11 32	11 17	10 59	10 37	10 09	9 27	■	■
15	13 47	13 41	13 34	13 26	13 18	13 09	12 59	12 47	12 34	12 19	12 00	11 37	11 06	10 17
16	15 04	14 59	14 54	14 49	14 43	14 37	14 30	14 22	14 13	14 03	13 52	13 38	13 22	13 02
17	16 21	16 18	16 15	16 12	16 09	16 06	16 02	15 57	15 52	15 47	15 41	15 34	15 26	15 16
18	17 38	17 38	17 37	17 36	17 35	17 34	17 33	17 32	17 31	17 30	17 28	17 27	17 25	17 22
19	18 56	18 58	18 59	19 00	19 02	19 04	19 06	19 08	19 10	19 13	19 16	19 19	19 23	19 27
20	20 15	20 19	20 22	20 26	20 30	20 35	20 39	20 45	20 51	20 58	21 06	21 15	21 25	21 38
21	21 35	21 40	21 46	21 52	21 59	22 06	22 15	22 24	22 34	22 46	22 59	23 16	23 36	.. ..
22	22 55	23 02	23 09	23 18	23 27	23 37	23 49	.. ..	.. ..	.. ..	.. ..	.. ..	.. ..	0 02
23	.. ..	.. ..	.. ..	.. ..	.. ..	.. ..	.. ..	0 02	0 17	0 35	0 56	1 25	2 08	□
24	0 10	0 19	0 28	0 38	0 49	1 01	1 16	1 32	1 52	2 16	2 49	3 46	□	□
25	1 17	1 26	1 36	1 47	1 59	2 12	2 28	2 46	3 08	3 35	4 16	□	□	□
26	2 14	2 22	2 32	2 42	2 53	3 06	3 20	3 37	3 57	4 22	4 56	5 55	□	□
27	2 58	3 06	3 14	3 23	3 33	3 44	3 56	4 10	4 26	4 45	5 09	5 40	6 30	□
28	3 34	3 40	3 47	3 54	4 02	4 11	4 20	4 31	4 43	4 57	5 13	5 33	5 58	6 32
29	4 02	4 07	4 12	4 18	4 24	4 30	4 37	4 45	4 53	5 03	5 14	5 27	5 43	6 01
30	4 26	4 30	4 33	4 37	4 41	4 45	4 50	4 55	5 00	5 07	5 14	5 22	5 31	5 42
31	4 47	4 49	4 51	4 53	4 55	4 57	5 00	5 03	5 06	5 09	5 13	5 17	5 22	5 27
Apr. 1	5 07	5 07	5 07	5 08	5 08	5 09	5 09	5 10	5 10	5 11	5 11	5 12	5 13	5 14
2	5 26	5 25	5 24	5 22	5 21	5 20	5 18	5 16	5 14	5 12	5 10	5 07	5 04	5 01

□ indicates Moon continuously above horizon.
■ indicates Moon continuously below horizon.
.. .. indicates phenomenon will occur the next day.

UNIVERSAL TIME FOR MERIDIAN OF GREENWICH

MOONRISE

Lat.	−55°	−50°	−45°	−40°	−35°	−30°	−20°	−10°	0°	+10°	+20°	+30°	+35°	+40°
	h m	h m	h m	h m	h m	h m	h m	h m	h m	h m	h m	h m	h m	h m
Apr. 1	17 14	17 15	17 16	17 17	17 18	17 19	17 20	17 21	17 22	17 24	17 25	17 26	17 27	17 28
2	17 19	17 26	17 32	17 37	17 41	17 44	17 51	17 56	18 02	18 07	18 13	18 19	18 23	18 27
3	17 26	17 39	17 49	17 57	18 05	18 11	18 22	18 32	18 42	18 51	19 01	19 13	19 20	19 28
4	17 34	17 53	18 08	18 20	18 31	18 40	18 56	19 10	19 24	19 37	19 52	20 08	20 18	20 29
5	17 46	18 11	18 31	18 47	19 01	19 13	19 34	19 52	20 09	20 26	20 44	21 05	21 18	21 32
6	18 03	18 36	19 00	19 20	19 36	19 51	20 15	20 37	20 57	21 17	21 38	22 03	22 18	22 35
7	18 29	19 09	19 38	20 00	20 19	20 35	21 02	21 26	21 48	22 10	22 34	23 01	23 18	23 37
8	19 12	19 55	20 25	20 49	21 09	21 26	21 54	22 19	22 42	23 04	23 29	23 58	.. ..	.. ..
9	20 13	20 55	21 25	21 48	22 07	22 23	22 51	23 15	23 37	23 59	.. ..	.. ..	0 15	0 34
10	21 32	22 08	22 34	22 54	23 11	23 26	23 51	.. ..	.. ..	.. ..	0 23	0 50	1 06	1 25
11	23 02	23 28	23 49	.. ..	.. ..	.. ..	.. ..	0 12	0 32	0 52	1 13	1 37	1 52	2 08
12	.. ..	.. ..	.. ..	0 05	0 19	0 31	0 52	1 09	1 26	1 42	2 00	2 20	2 31	2 44
13	0 35	0 53	1 07	1 19	1 29	1 38	1 53	2 06	2 18	2 30	2 43	2 57	3 06	3 15
14	2 09	2 19	2 27	2 34	2 40	2 45	2 54	3 01	3 09	3 16	3 23	3 32	3 37	3 42
15	3 43	3 46	3 48	3 50	3 51	3 53	3 55	3 57	3 59	4 01	4 03	4 05	4 06	4 08
16	5 20	5 14	5 10	5 07	5 04	5 01	4 57	4 53	4 50	4 46	4 42	4 38	4 36	4 33
17	6 59	6 45	6 35	6 26	6 19	6 12	6 01	5 51	5 42	5 33	5 24	5 14	5 07	5 01
18	8 41	8 19	8 02	7 47	7 36	7 25	7 08	6 52	6 38	6 24	6 09	5 53	5 43	5 32
19	10 25	9 52	9 28	9 09	8 53	8 40	8 16	7 56	7 38	7 19	7 00	6 37	6 24	6 09
20	12 00	11 19	10 50	10 27	10 08	9 52	9 25	9 02	8 40	8 18	7 55	7 29	7 13	6 55
21	13 16	12 31	12 00	11 36	11 16	10 59	10 30	10 06	9 43	9 20	8 56	8 27	8 10	7 51
22	14 05	13 25	12 56	12 33	12 14	11 57	11 30	11 06	10 44	10 22	9 58	9 30	9 14	8 54
23	14 35	14 02	13 37	13 17	13 01	12 46	12 22	12 00	11 41	11 21	10 59	10 34	10 20	10 03
24	14 52	14 27	14 08	13 52	13 39	13 27	13 07	12 49	12 32	12 16	11 58	11 37	11 25	11 11
25	15 04	14 46	14 32	14 20	14 10	14 01	13 46	13 32	13 19	13 07	12 53	12 37	12 28	12 17

MOONSET

Lat.	−55°	−50°	−45°	−40°	−35°	−30°	−20°	−10°	0°	+10°	+20°	+30°	+35°	+40°
	h m	h m	h m	h m	h m	h m	h m	h m	h m	h m	h m	h m	h m	h m
Apr. 1	4 54	4 55	4 56	4 57	4 58	4 59	5 00	5 01	5 02	5 03	5 04	5 05	5 06	5 07
2	6 08	6 03	6 00	5 57	5 54	5 52	5 48	5 44	5 41	5 38	5 35	5 31	5 28	5 26
3	7 22	7 12	7 03	6 56	6 50	6 45	6 36	6 28	6 21	6 13	6 05	5 57	5 51	5 46
4	8 39	8 22	8 08	7 57	7 48	7 39	7 25	7 13	7 02	6 50	6 38	6 24	6 16	6 07
5	9 57	9 33	9 14	8 59	8 47	8 36	8 17	8 00	7 45	7 30	7 14	6 55	6 44	6 32
6	11 17	10 45	10 21	10 03	9 47	9 33	9 10	8 50	8 31	8 13	7 53	7 30	7 17	7 02
7	12 34	11 55	11 27	11 05	10 47	10 31	10 05	9 42	9 21	9 00	8 37	8 11	7 56	7 38
8	13 42	12 59	12 29	12 05	11 45	11 29	11 00	10 36	10 14	9 51	9 27	8 58	8 42	8 22
9	14 35	13 53	13 23	12 59	12 40	12 23	11 55	11 31	11 08	10 45	10 21	9 53	9 36	9 16
10	15 10	14 34	14 07	13 46	13 28	13 13	12 47	12 24	12 03	11 42	11 19	10 53	10 37	10 18
11	15 33	15 05	14 43	14 25	14 11	13 58	13 35	13 16	12 58	12 39	12 20	11 57	11 43	11 28
12	15 48	15 28	15 12	14 59	14 47	14 37	14 20	14 05	13 51	13 36	13 21	13 03	12 52	12 40
13	15 59	15 46	15 36	15 27	15 20	15 13	15 02	14 52	14 42	14 32	14 22	14 10	14 03	13 55
14	16 08	16 02	15 58	15 54	15 50	15 47	15 42	15 37	15 32	15 28	15 23	15 17	15 14	15 10
15	16 16	16 17	16 18	16 19	16 19	16 20	16 21	16 22	16 23	16 23	16 24	16 25	16 26	16 26
16	16 24	16 33	16 39	16 45	16 50	16 54	17 01	17 08	17 14	17 20	17 27	17 35	17 39	17 44
17	16 34	16 50	17 03	17 14	17 23	17 31	17 45	17 57	18 08	18 20	18 32	18 47	18 55	19 04
18	16 48	17 13	17 32	17 47	18 01	18 12	18 32	18 50	19 06	19 22	19 40	20 00	20 12	20 26
19	17 10	17 43	18 09	18 29	18 46	19 00	19 25	19 47	20 07	20 27	20 49	21 14	21 29	21 47
20	17 45	18 26	18 56	19 20	19 39	19 55	20 23	20 48	21 10	21 32	21 57	22 25	22 41	23 01
21	18 40	19 24	19 56	20 20	20 40	20 56	21 25	21 50	22 12	22 35	22 59	23 28	23 44	.. ..
22	19 56	20 35	21 04	21 26	21 45	22 01	22 27	22 50	23 12	23 33	23 55	.. ..	.. ..	0 04
23	21 21	21 53	22 17	22 35	22 51	23 05	23 28	23 48	.. ..	.. ..	.. ..	0 21	0 37	0 55
24	22 47	23 11	23 29	23 43	23 56	.. ..	.. ..	.. ..	0 06	0 24	0 44	1 06	1 19	1 34
25	.. ..	.. ..	.. ..	.. ..	.. ..	0 06	0 25	0 41	0 55	1 10	1 26	1 43	1 54	2 05

.. .. indicates phenomenon will occur the next day.

UNIVERSAL TIME FOR MERIDIAN OF GREENWICH

MOONRISE

Lat.	+40°	+42°	+44°	+46°	+48°	+50°	+52°	+54°	+56°	+58°	+60°	+62°	+64°	+66°
	h m	h m	h m	h m	h m	h m	h m	h m	h m	h m	h m	h m	h m	h m
Apr. 1	17 28	17 28	17 29	17 29	17 30	17 30	17 31	17 32	17 32	17 33	17 34	17 35	17 36	17 38
2	18 27	18 29	18 31	18 34	18 36	18 39	18 41	18 45	18 48	18 52	18 56	19 01	19 07	19 14
3	19 28	19 31	19 35	19 39	19 43	19 48	19 53	19 59	20 05	20 12	20 20	20 30	20 41	20 54
4	20 29	20 34	20 39	20 45	20 52	20 58	21 06	21 15	21 24	21 35	21 48	22 03	22 21	22 45
5	21 32	21 38	21 45	21 53	22 01	22 11	22 21	22 32	22 46	23 01	23 20	23 43	.. ..	.. ..
6	22 35	22 43	22 52	23 01	23 11	23 22	23 35	23 50	.. ..	.. ..	.. ..	.. ..	0 15	1 07
7	23 37	23 46	23 55	.. ..	.. ..	.. ..	.. ..	.. ..	0 08	0 29	0 56	1 35	■	■
8	.. ..	.. ..	.. ..	0 06	0 17	0 30	0 45	1 03	1 24	1 50	2 28	■	■	■
9	0 34	0 43	0 53	1 04	1 16	1 30	1 45	2 03	2 25	2 53	3 35	■	■	■
10	1 25	1 34	1 43	1 53	2 04	2 17	2 31	2 48	3 07	3 32	4 04	4 59	■	■
11	2 08	2 16	2 24	2 32	2 42	2 53	3 05	3 18	3 34	3 52	4 14	4 44	5 27	■
12	2 44	2 50	2 57	3 04	3 11	3 19	3 28	3 38	3 50	4 03	4 18	4 36	4 58	5 27
13	3 15	3 19	3 24	3 29	3 34	3 40	3 46	3 53	4 00	4 09	4 19	4 30	4 43	4 58
14	3 42	3 45	3 48	3 50	3 53	3 57	4 00	4 04	4 08	4 13	4 18	4 24	4 31	4 39
15	4 08	4 09	4 09	4 10	4 11	4 12	4 13	4 14	4 15	4 16	4 18	4 19	4 21	4 23
16	4 33	4 32	4 31	4 30	4 28	4 27	4 25	4 23	4 21	4 19	4 17	4 14	4 11	4 08
17	5 01	4 58	4 54	4 51	4 47	4 43	4 39	4 34	4 29	4 23	4 17	4 09	4 01	3 51
18	5 32	5 27	5 22	5 16	5 10	5 03	4 56	4 48	4 39	4 29	4 18	4 04	3 48	3 29
19	6 09	6 02	5 55	5 47	5 39	5 30	5 19	5 08	4 54	4 39	4 21	3 58	3 30	2 48
20	6 55	6 47	6 38	6 28	6 18	6 06	5 53	5 37	5 20	4 58	4 30	3 51	□	□
21	7 51	7 42	7 32	7 21	7 10	6 56	6 41	6 24	6 03	5 36	4 58	□	□	□
22	8 54	8 46	8 36	8 26	8 14	8 01	7 47	7 29	7 09	6 43	6 07	4 44	□	□
23	10 03	9 55	9 47	9 37	9 27	9 16	9 03	8 49	8 32	8 11	7 45	7 09	5 43	□
24	11 11	11 05	10 58	10 51	10 43	10 34	10 24	10 13	10 00	9 45	9 28	9 06	8 37	7 54
25	12 17	12 13	12 08	12 02	11 56	11 50	11 43	11 35	11 26	11 16	11 04	10 50	10 34	10 13

MOONSET

Lat.	+40°	+42°	+44°	+46°	+48°	+50°	+52°	+54°	+56°	+58°	+60°	+62°	+64°	+66°
	h m	h m	h m	h m	h m	h m	h m	h m	h m	h m	h m	h m	h m	h m
Apr. 1	5 07	5 07	5 07	5 08	5 08	5 09	5 09	5 10	5 10	5 11	5 11	5 12	5 13	5 14
2	5 26	5 25	5 24	5 22	5 21	5 20	5 18	5 16	5 14	5 12	5 10	5 07	5 04	5 01
3	5 46	5 43	5 40	5 38	5 34	5 31	5 27	5 23	5 19	5 14	5 09	5 02	4 55	4 47
4	6 07	6 03	5 59	5 54	5 50	5 44	5 38	5 32	5 25	5 17	5 08	4 57	4 45	4 30
5	6 32	6 27	6 21	6 14	6 08	6 00	5 52	5 43	5 32	5 21	5 07	4 51	4 32	4 08
6	7 02	6 55	6 47	6 39	6 31	6 21	6 10	5 58	5 44	5 28	5 09	4 45	4 12	3 20
7	7 38	7 30	7 21	7 11	7 01	6 49	6 36	6 21	6 03	5 42	5 14	4 35	■	■
8	8 22	8 13	8 04	7 53	7 41	7 28	7 13	6 56	6 35	6 08	5 30	■	■	■
9	9 16	9 07	8 57	8 46	8 34	8 21	8 06	7 47	7 26	6 57	6 16	■	■	■
10	10 18	10 10	10 01	9 51	9 40	9 28	9 14	8 57	8 38	8 14	7 42	6 47	■	■
11	11 28	11 20	11 13	11 04	10 55	10 45	10 34	10 21	10 06	9 48	9 26	8 57	8 15	■
12	12 40	12 35	12 29	12 23	12 16	12 08	12 00	11 51	11 40	11 28	11 14	10 56	10 35	10 07
13	13 55	13 51	13 47	13 43	13 39	13 34	13 28	13 22	13 16	13 08	13 00	12 50	12 38	12 24
14	15 10	15 08	15 06	15 04	15 02	15 00	14 58	14 55	14 52	14 48	14 45	14 40	14 35	14 29
15	16 26	16 26	16 27	16 27	16 27	16 28	16 28	16 28	16 29	16 29	16 30	16 30	16 31	16 32
16	17 44	17 46	17 49	17 51	17 54	17 57	18 00	18 04	18 08	18 12	18 18	18 23	18 30	18 38
17	19 04	19 09	19 13	19 18	19 24	19 29	19 36	19 43	19 51	20 00	20 10	20 22	20 37	20 54
18	20 26	20 33	20 39	20 46	20 54	21 03	21 13	21 24	21 36	21 51	22 08	22 30	22 58	23 39
19	21 47	21 55	22 03	22 12	22 23	22 34	22 47	23 02	23 19	23 41	.. ..	.. ..	□	□
20	23 01	23 10	23 19	23 30	23 41	23 55	.. ..	.. ..	.. ..	.. ..	0 08	0 47	□	□
21	.. ..	.. ..	.. ..	.. ..	.. ..	.. ..	0 10	0 27	0 48	1 15	1 52	□	□	□
22	0 04	0 13	0 22	0 33	0 45	0 58	1 13	1 30	1 51	2 17	2 53	4 16	□	□
23	0 55	1 03	1 11	1 21	1 31	1 43	1 56	2 11	2 28	2 49	3 15	3 52	5 19	□
24	1 34	1 41	1 48	1 56	2 04	2 14	2 24	2 36	2 49	3 04	3 23	3 45	4 15	4 59
25	2 05	2 11	2 16	2 22	2 29	2 36	2 43	2 52	3 02	3 13	3 25	3 40	3 58	4 19

□ indicates Moon continuously above horizon.
■ indicates Moon continuously below horizon.
.. .. indicates phenomenon will occur the next day.

UNIVERSAL TIME FOR MERIDIAN OF GREENWICH

MOONRISE

Lat.	−55°	−50°	−45°	−40°	−35°	−30°	−20°	−10°	0°	+10°	+20°	+30°	+35°	+40°
	h m	h m	h m	h m	h m	h m	h m	h m	h m	h m	h m	h m	h m	h m
Apr. 24	14 52	14 27	14 08	13 52	13 39	13 27	13 07	12 49	12 32	12 16	11 58	11 37	11 25	11 11
25	15 04	14 46	14 32	14 20	14 10	14 01	13 46	13 32	13 19	13 07	12 53	12 37	12 28	12 17
26	15 12	15 00	14 51	14 43	14 37	14 31	14 20	14 11	14 03	13 54	13 45	13 34	13 28	13 21
27	15 18	15 13	15 08	15 04	15 01	14 58	14 52	14 48	14 43	14 39	14 34	14 29	14 25	14 22
28	15 24	15 24	15 24	15 23	15 23	15 23	15 23	15 22	15 22	15 22	15 22	15 22	15 22	15 21
29	15 30	15 35	15 39	15 43	15 46	15 48	15 53	15 57	16 01	16 05	16 09	16 14	16 17	16 20
30	15 36	15 47	15 55	16 03	16 09	16 15	16 24	16 33	16 41	16 49	16 58	17 08	17 14	17 20
May 1	15 44	16 01	16 14	16 25	16 35	16 43	16 57	17 10	17 22	17 34	17 47	18 02	18 11	18 21
2	15 55	16 18	16 36	16 51	17 04	17 15	17 34	17 51	18 06	18 22	18 39	18 59	19 11	19 24
3	16 10	16 40	17 03	17 22	17 37	17 51	18 14	18 35	18 54	19 13	19 33	19 57	20 11	20 27
4	16 34	17 11	17 38	18 00	18 18	18 33	19 00	19 23	19 44	20 05	20 29	20 55	21 11	21 30
5	17 11	17 53	18 23	18 46	19 06	19 22	19 50	20 15	20 37	21 00	21 24	21 52	22 09	22 29
6	18 06	18 48	19 18	19 42	20 01	20 18	20 46	21 10	21 32	21 54	22 18	22 46	23 02	23 21
7	19 20	19 57	20 24	20 45	21 03	21 18	21 44	22 06	22 26	22 47	23 09	23 34	23 49	.. ..
8	20 44	21 14	21 36	21 53	22 08	22 21	22 43	23 02	23 19	23 37	23 56	.. ..	.. ..	0 06
9	22 14	22 35	22 51	23 04	23 16	23 26	23 42	23 57	.. ..	.. ..	.. ..	0 17	0 30	0 44
10	23 44	23 57	.. ..	.. ..	.. ..	.. ..	.. ..	.. ..	0 11	0 24	0 39	0 55	1 05	1 15
11	.. ..	.. ..	0 08	0 16	0 24	0 30	0 41	0 51	1 00	1 09	1 19	1 30	1 36	1 43
12	1 14	1 20	1 25	1 29	1 32	1 35	1 40	1 44	1 48	1 53	1 57	2 02	2 05	2 08
13	2 46	2 45	2 43	2 42	2 41	2 41	2 39	2 38	2 37	2 36	2 35	2 34	2 33	2 33
14	4 21	4 12	4 04	3 58	3 53	3 48	3 40	3 34	3 27	3 21	3 15	3 07	3 03	2 58
15	6 00	5 42	5 28	5 17	5 07	4 59	4 45	4 32	4 21	4 09	3 57	3 43	3 36	3 27
16	7 42	7 15	6 55	6 38	6 24	6 12	5 52	5 34	5 18	5 02	4 44	4 25	4 13	4 00
17	9 23	8 47	8 20	7 59	7 41	7 26	7 01	6 39	6 19	5 59	5 38	5 13	4 59	4 42
18	10 51	10 08	9 38	9 14	8 55	8 38	8 10	7 46	7 23	7 01	6 37	6 09	5 53	5 34

MOONSET

Lat.	−55°	−50°	−45°	−40°	−35°	−30°	−20°	−10°	0°	+10°	+20°	+30°	+35°	+40°
	h m	h m	h m	h m	h m	h m	h m	h m	h m	h m	h m	h m	h m	h m
Apr. 24	22 47	23 11	23 29	23 43	23 56	.. ..	.. ..	.. ..	0 06	0 24	0 44	1 06	1 19	1 34
25	.. ..	.. ..	.. ..	.. ..	.. ..	0 06	0 25	0 41	0 55	1 10	1 26	1 43	1 54	2 05
26	0 09	0 25	0 38	0 48	0 57	1 05	1 18	1 30	1 40	1 51	2 02	2 15	2 23	2 31
27	1 27	1 37	1 44	1 51	1 56	2 01	2 09	2 16	2 22	2 29	2 36	2 43	2 48	2 53
28	2 43	2 46	2 49	2 51	2 53	2 54	2 57	3 00	3 02	3 04	3 07	3 09	3 11	3 13
29	3 56	3 54	3 52	3 50	3 48	3 47	3 45	3 43	3 41	3 39	3 37	3 35	3 33	3 32
30	5 10	5 02	4 55	4 49	4 44	4 40	4 33	4 26	4 20	4 14	4 08	4 00	3 56	3 51
May 1	6 26	6 11	5 59	5 50	5 41	5 34	5 22	5 11	5 00	4 50	4 40	4 27	4 20	4 13
2	7 44	7 22	7 05	6 51	6 40	6 30	6 12	5 57	5 43	5 29	5 14	4 57	4 48	4 36
3	9 03	8 34	8 12	7 54	7 40	7 27	7 05	6 46	6 29	6 11	5 53	5 31	5 19	5 04
4	10 22	9 45	9 19	8 58	8 40	8 25	8 00	7 38	7 18	6 57	6 35	6 10	5 56	5 39
5	11 33	10 52	10 22	9 59	9 40	9 23	8 55	8 32	8 09	7 47	7 24	6 56	6 40	6 21
6	12 31	11 48	11 18	10 55	10 35	10 19	9 50	9 26	9 03	8 41	8 16	7 48	7 31	7 12
7	13 11	12 33	12 05	11 43	11 25	11 09	10 43	10 20	9 58	9 36	9 13	8 46	8 30	8 11
8	13 37	13 06	12 43	12 24	12 09	11 55	11 32	11 11	10 52	10 33	10 12	9 48	9 34	9 17
9	13 54	13 31	13 13	12 59	12 46	12 35	12 16	12 00	11 44	11 28	11 11	10 52	10 40	10 27
10	14 06	13 51	13 38	13 28	13 19	13 11	12 58	12 46	12 34	12 23	12 11	11 56	11 48	11 39
11	14 15	14 07	14 00	13 54	13 49	13 45	13 37	13 30	13 23	13 16	13 09	13 01	12 56	12 51
12	14 23	14 22	14 20	14 19	14 17	14 16	14 15	14 13	14 11	14 10	14 08	14 06	14 05	14 03
13	14 31	14 36	14 40	14 43	14 46	14 49	14 53	14 57	15 01	15 04	15 08	15 13	15 15	15 18
14	14 40	14 52	15 02	15 10	15 17	15 23	15 34	15 43	15 52	16 01	16 11	16 21	16 28	16 35
15	14 52	15 12	15 28	15 41	15 52	16 02	16 18	16 33	16 47	17 01	17 16	17 33	17 44	17 55
16	15 10	15 38	16 00	16 18	16 33	16 46	17 09	17 28	17 46	18 05	18 25	18 48	19 01	19 17
17	15 37	16 15	16 43	17 05	17 23	17 38	18 05	18 28	18 49	19 11	19 34	20 01	20 17	20 35
18	16 24	17 07	17 38	18 02	18 21	18 38	19 07	19 31	19 54	20 17	20 41	21 09	21 26	21 46

.. .. indicates phenomenon will occur the next day.

UNIVERSAL TIME FOR MERIDIAN OF GREENWICH

MOONRISE

Lat.	+40°	+42°	+44°	+46°	+48°	+50°	+52°	+54°	+56°	+58°	+60°	+62°	+64°	+66°
	h m	h m	h m	h m	h m	h m	h m	h m	h m	h m	h m	h m	h m	h m
Apr. 24	11 11	11 05	10 58	10 51	10 43	10 34	10 24	10 13	10 00	9 45	9 28	9 06	8 37	7 54
25	12 17	12 13	12 08	12 02	11 56	11 50	11 43	11 35	11 26	11 16	11 04	10 50	10 34	10 13
26	13 21	13 18	13 15	13 11	13 07	13 03	12 58	12 53	12 47	12 41	12 34	12 26	12 16	12 04
27	14 22	14 20	14 19	14 17	14 15	14 13	14 11	14 08	14 05	14 02	13 59	13 55	13 50	13 45
28	15 21	15 21	15 21	15 21	15 21	15 21	15 21	15 21	15 21	15 21	15 21	15 20	15 20	15 20
29	16 20	16 22	16 23	16 25	16 27	16 29	16 31	16 33	16 36	16 39	16 42	16 46	16 50	16 55
30	17 20	17 23	17 26	17 30	17 33	17 37	17 42	17 47	17 52	17 58	18 05	18 13	18 22	18 33
May 1	18 21	18 26	18 31	18 36	18 41	18 48	18 54	19 02	19 10	19 20	19 31	19 44	20 00	20 19
2	19 24	19 30	19 36	19 43	19 51	19 59	20 09	20 19	20 31	20 45	21 02	21 22	21 48	22 24
3	20 27	20 35	20 43	20 52	21 01	21 12	21 24	21 37	21 53	22 12	22 36	23 08	.. ..	▬
4	21 30	21 38	21 48	21 58	22 09	22 21	22 36	22 52	23 12	23 36	.. ..	.. ..	0 02	▬
5	22 29	22 38	22 48	22 58	23 10	23 24	23 39	23 57	.. ..	.. ..	0 10	1 10	▬	▬
6	23 21	23 30	23 40	23 50	.. ..	.. ..	.. ..	.. ..	0 18	0 46	1 26	▬	▬	▬
7	.. ..	.. ..	.. ..	.. ..	0 01	0 14	0 29	0 46	1 06	1 32	2 06	3 13	▬	▬
8	0 06	0 14	0 23	0 32	0 42	0 53	1 06	1 20	1 36	1 56	2 21	2 55	3 54	▬
9	0 44	0 50	0 57	1 05	1 13	1 22	1 32	1 43	1 55	2 10	2 27	2 48	3 14	3 51
10	1 15	1 20	1 25	1 31	1 37	1 43	1 51	1 59	2 07	2 17	2 29	2 42	2 57	3 16
11	1 43	1 46	1 49	1 53	1 57	2 01	2 06	2 10	2 16	2 22	2 29	2 37	2 46	2 56
12	2 08	2 09	2 11	2 13	2 14	2 16	2 18	2 20	2 23	2 26	2 29	2 32	2 36	2 40
13	2 33	2 32	2 32	2 32	2 31	2 31	2 30	2 30	2 29	2 29	2 28	2 27	2 27	2 26
14	2 58	2 56	2 54	2 51	2 49	2 46	2 43	2 40	2 36	2 32	2 28	2 23	2 17	2 11
15	3 27	3 23	3 18	3 14	3 09	3 04	2 58	2 52	2 45	2 37	2 28	2 18	2 07	1 53
16	4 00	3 55	3 49	3 42	3 35	3 27	3 18	3 08	2 57	2 45	2 31	2 14	1 53	1 26
17	4 42	4 35	4 27	4 18	4 08	3 58	3 46	3 33	3 17	2 59	2 37	2 08	1 27	▭
18	5 34	5 25	5 16	5 06	4 55	4 42	4 28	4 11	3 52	3 27	2 55	2 01	▭	▭

MOONSET

Lat.	+40°	+42°	+44°	+46°	+48°	+50°	+52°	+54°	+56°	+58°	+60°	+62°	+64°	+66°
	h m	h m	h m	h m	h m	h m	h m	h m	h m	h m	h m	h m	h m	h m
Apr. 24	1 34	1 41	1 48	1 56	2 04	2 14	2 24	2 36	2 49	3 04	3 23	3 45	4 15	4 59
25	2 05	2 11	2 16	2 22	2 29	2 36	2 43	2 52	3 02	3 13	3 25	3 40	3 58	4 19
26	2 31	2 35	2 39	2 43	2 47	2 52	2 58	3 03	3 10	3 17	3 26	3 35	3 46	3 59
27	2 53	2 55	2 57	3 00	3 03	3 06	3 09	3 12	3 16	3 20	3 25	3 30	3 36	3 43
28	3 13	3 14	3 14	3 15	3 16	3 17	3 18	3 19	3 21	3 22	3 24	3 26	3 28	3 30
29	3 32	3 31	3 31	3 30	3 29	3 28	3 27	3 26	3 25	3 24	3 22	3 21	3 19	3 17
30	3 51	3 49	3 47	3 45	3 42	3 40	3 37	3 33	3 30	3 26	3 21	3 16	3 10	3 04
May 1	4 13	4 09	4 05	4 01	3 57	3 52	3 47	3 41	3 35	3 28	3 20	3 11	3 01	2 49
2	4 36	4 31	4 26	4 20	4 14	4 07	4 00	3 52	3 43	3 32	3 20	3 06	2 50	2 29
3	5 04	4 58	4 51	4 44	4 36	4 27	4 17	4 06	3 53	3 39	3 21	3 01	2 34	1 57
4	5 39	5 31	5 23	5 14	5 04	4 53	4 40	4 26	4 10	3 50	3 26	2 53	1 59	▬
5	6 21	6 12	6 03	5 52	5 41	5 28	5 14	4 57	4 37	4 12	3 39	2 39	▬	▬
6	7 12	7 03	6 53	6 42	6 30	6 17	6 02	5 44	5 22	4 54	4 14	▬	▬	▬
7	8 11	8 03	7 53	7 43	7 32	7 19	7 05	6 48	6 28	6 03	5 28	4 22	▬	▬
8	9 17	9 10	9 02	8 53	8 43	8 32	8 20	8 06	7 50	7 30	7 06	6 33	5 35	▬
9	10 27	10 21	10 15	10 08	10 00	9 52	9 42	9 32	9 20	9 06	8 50	8 30	8 04	7 28
10	11 39	11 34	11 30	11 25	11 20	11 14	11 07	11 00	10 52	10 43	10 33	10 21	10 06	9 49
11	12 51	12 48	12 46	12 43	12 40	12 37	12 33	12 29	12 25	12 20	12 14	12 08	12 00	11 51
12	14 03	14 03	14 02	14 02	14 01	14 00	13 59	13 58	13 57	13 56	13 55	13 53	13 51	13 49
13	15 18	15 19	15 21	15 22	15 24	15 25	15 27	15 29	15 32	15 34	15 37	15 40	15 44	15 48
14	16 35	16 38	16 42	16 45	16 50	16 54	16 59	17 04	17 10	17 17	17 24	17 33	17 43	17 55
15	17 55	18 00	18 06	18 12	18 19	18 26	18 34	18 43	18 53	19 04	19 18	19 34	19 53	20 18
16	19 17	19 24	19 31	19 40	19 49	19 59	20 10	20 23	20 38	20 55	21 17	21 45	22 25	▭
17	20 35	20 44	20 53	21 03	21 14	21 26	21 40	21 57	22 16	22 40	23 12	.. ..	▭	▭
18	21 46	21 55	22 04	22 15	22 27	22 40	22 55	23 12	23 34	.. ..	.. ..	0 05	▭	▭

▭ indicates Moon continuously above horizon.
▬ indicates Moon continuously below horizon.
.. .. indicates phenomenon will occur the next day.

UNIVERSAL TIME FOR MERIDIAN OF GREENWICH

MOONRISE

Lat.	−55°	−50°	−45°	−40°	−35°	−30°	−20°	−10°	0°	+10°	+20°	+30°	+35°	+40°
	h m	h m	h m	h m	h m	h m	h m	h m	h m	h m	h m	h m	h m	h m
May 17	9 23	8 47	8 20	7 59	7 41	7 26	7 01	6 39	6 19	5 59	5 38	5 13	4 59	4 42
18	10 51	10 08	9 38	9 14	8 55	8 38	8 10	7 46	7 23	7 01	6 37	6 09	5 53	5 34
19	11 55	11 12	10 42	10 18	9 59	9 42	9 14	8 50	8 27	8 04	7 40	7 12	6 55	6 36
20	12 33	11 57	11 31	11 10	10 52	10 37	10 11	9 49	9 28	9 07	8 44	8 18	8 03	7 45
21	12 56	12 28	12 07	11 50	11 35	11 22	11 00	10 41	10 23	10 05	9 46	9 24	9 11	8 56
22	13 10	12 50	12 34	12 21	12 10	12 00	11 43	11 28	11 13	10 59	10 44	10 27	10 16	10 05
23	13 20	13 06	12 55	12 46	12 38	12 32	12 20	12 09	11 59	11 49	11 38	11 26	11 19	11 11
24	13 27	13 20	13 13	13 08	13 04	13 00	12 53	12 47	12 41	12 35	12 29	12 22	12 18	12 13
25	13 33	13 31	13 30	13 28	13 27	13 26	13 24	13 22	13 21	13 19	13 18	13 16	13 15	13 14
26	13 39	13 42	13 45	13 48	13 50	13 51	13 54	13 57	14 00	14 03	14 05	14 09	14 11	14 13
27	13 45	13 54	14 01	14 07	14 13	14 17	14 25	14 32	14 39	14 46	14 53	15 02	15 07	15 12
28	13 53	14 07	14 19	14 29	14 37	14 45	14 58	15 09	15 20	15 31	15 42	15 56	16 04	16 13
29	14 02	14 23	14 40	14 54	15 05	15 16	15 33	15 49	16 03	16 18	16 34	16 52	17 03	17 15
30	14 16	14 44	15 06	15 23	15 38	15 51	16 12	16 32	16 50	17 08	17 27	17 50	18 03	18 18
31	14 37	15 12	15 38	15 59	16 16	16 31	16 57	17 19	17 39	18 00	18 22	18 48	19 04	19 22
June 1	15 10	15 51	16 20	16 43	17 02	17 19	17 46	18 10	18 32	18 55	19 19	19 47	20 03	20 23
2	16 01	16 43	17 13	17 37	17 56	18 13	18 41	19 05	19 27	19 50	20 14	20 42	20 58	21 18
3	17 10	17 49	18 16	18 38	18 57	19 12	19 38	20 01	20 22	20 43	21 06	21 32	21 47	22 05
4	18 32	19 04	19 27	19 46	20 01	20 15	20 38	20 58	21 16	21 34	21 54	22 17	22 30	22 45
5	20 00	20 24	20 41	20 56	21 08	21 19	21 37	21 53	22 08	22 22	22 38	22 56	23 06	23 18
6	21 29	21 45	21 57	22 07	22 15	22 23	22 35	22 46	22 57	23 07	23 18	23 31	23 38	23 46
7	22 58	23 06	23 12	23 17	23 22	23 26	23 33	23 39	23 44	23 50	23 56	.. ..	.. ..	.. ..
8	.. ..	.. ..	.. ..	.. ..	.. ..	.. ..	.. ..	.. ..	.. ..	.. ..	.. ..	0 03	0 07	0 11
9	0 26	0 27	0 28	0 28	0 29	0 29	0 30	0 31	0 31	0 32	0 33	0 34	0 34	0 35
10	1 56	1 50	1 45	1 41	1 37	1 34	1 28	1 24	1 19	1 15	1 10	1 05	1 02	0 59

MOONSET

Lat.	−55°	−50°	−45°	−40°	−35°	−30°	−20°	−10°	0°	+10°	+20°	+30°	+35°	+40°
	h m	h m	h m	h m	h m	h m	h m	h m	h m	h m	h m	h m	h m	h m
May 17	15 37	16 15	16 43	17 05	17 23	17 38	18 05	18 28	18 49	19 11	19 34	20 01	20 17	20 35
18	16 24	17 07	17 38	18 02	18 21	18 38	19 07	19 31	19 54	20 17	20 41	21 09	21 26	21 46
19	17 33	18 14	18 44	19 08	19 27	19 43	20 11	20 35	20 57	21 19	21 42	22 09	22 25	22 44
20	18 57	19 32	19 58	20 18	20 35	20 50	21 14	21 35	21 55	22 14	22 35	22 59	23 13	23 29
21	20 26	20 53	21 13	21 29	21 43	21 54	22 14	22 32	22 48	23 04	23 21	23 40	23 52	.. ..
22	21 52	22 10	22 25	22 37	22 47	22 56	23 11	23 24	23 36	23 48	.. ..	.. ..	.. ..	0 05
23	23 13	23 24	23 34	23 41	23 48	23 53	.. ..	.. ..	.. ..	.. ..	0 01	0 15	0 23	0 33
24	.. ..	.. ..	.. ..	.. ..	.. ..	.. ..	0 03	0 12	0 19	0 27	0 36	0 45	0 50	0 56
25	0 30	0 35	0 39	0 43	0 46	0 48	0 53	0 57	1 00	1 04	1 08	1 12	1 15	1 17
26	1 44	1 43	1 43	1 42	1 42	1 41	1 41	1 40	1 40	1 39	1 38	1 38	1 37	1 37
27	2 58	2 51	2 46	2 41	2 38	2 34	2 28	2 23	2 19	2 14	2 09	2 03	2 00	1 56
28	4 13	4 00	3 50	3 41	3 34	3 28	3 17	3 07	2 59	2 50	2 40	2 30	2 24	2 17
29	5 29	5 10	4 55	4 42	4 32	4 23	4 07	3 53	3 40	3 28	3 14	2 59	2 50	2 40
30	6 49	6 22	6 02	5 45	5 32	5 20	4 59	4 42	4 25	4 09	3 51	3 31	3 20	3 07
31	8 08	7 34	7 09	6 49	6 32	6 18	5 54	5 33	5 13	4 54	4 33	4 09	3 55	3 39
June 1	9 23	8 43	8 14	7 51	7 33	7 17	6 50	6 26	6 05	5 43	5 20	4 53	4 37	4 19
2	10 26	9 43	9 13	8 50	8 30	8 13	7 45	7 21	6 59	6 36	6 12	5 44	5 27	5 07
3	11 11	10 32	10 03	9 41	9 22	9 06	8 39	8 16	7 54	7 32	7 08	6 41	6 24	6 05
4	11 41	11 08	10 44	10 24	10 08	9 54	9 29	9 08	8 48	8 28	8 07	7 42	7 27	7 10
5	12 00	11 35	11 16	11 00	10 47	10 35	10 15	9 58	9 41	9 24	9 06	8 46	8 33	8 19
6	12 13	11 56	11 42	11 31	11 21	11 12	10 57	10 44	10 31	10 19	10 05	9 49	9 40	9 30
7	12 23	12 13	12 04	11 57	11 51	11 46	11 36	11 28	11 20	11 12	11 03	10 53	10 47	10 40
8	12 32	12 28	12 24	12 22	12 19	12 17	12 13	12 10	12 07	12 03	12 00	11 56	11 54	11 51
9	12 39	12 42	12 44	12 45	12 47	12 48	12 50	12 52	12 54	12 56	12 58	13 00	13 01	13 02
10	12 47	12 57	13 04	13 10	13 15	13 20	13 28	13 36	13 43	13 50	13 57	14 05	14 10	14 16

.. .. indicates phenomenon will occur the next day.

UNIVERSAL TIME FOR MERIDIAN OF GREENWICH

MOONRISE

Lat.	+40°	+42°	+44°	+46°	+48°	+50°	+52°	+54°	+56°	+58°	+60°	+62°	+64°	+66°
	h m	h m	h m	h m	h m	h m	h m	h m	h m	h m	h m	h m	h m	h m
May 17	4 42	4 35	4 27	4 18	4 08	3 58	3 46	3 33	3 17	2 59	2 37	2 08	1 27	□
18	5 34	5 25	5 16	5 06	4 55	4 42	4 28	4 11	3 52	3 27	2 55	2 01	□	□
19	6 36	6 27	6 17	6 07	5 55	5 42	5 27	5 09	4 48	4 22	3 44	□	□	□
20	7 45	7 36	7 28	7 18	7 07	6 55	6 42	6 26	6 08	5 45	5 15	4 29	□	□
21	8 56	8 49	8 41	8 33	8 24	8 15	8 04	7 51	7 37	7 20	7 00	6 34	5 57	4 36
22	10 05	9 59	9 54	9 48	9 41	9 34	9 26	9 17	9 07	8 55	8 41	8 25	8 05	7 40
23	11 11	11 07	11 03	10 59	10 54	10 49	10 44	10 38	10 31	10 24	10 15	10 05	9 54	9 40
24	12 13	12 11	12 09	12 07	12 04	12 01	11 58	11 55	11 51	11 47	11 43	11 37	11 31	11 24
25	13 14	13 13	13 12	13 12	13 11	13 10	13 10	13 09	13 08	13 07	13 06	13 04	13 03	13 01
26	14 13	14 14	14 15	14 16	14 17	14 18	14 20	14 21	14 23	14 25	14 27	14 30	14 32	14 36
27	15 12	15 15	15 17	15 20	15 23	15 26	15 30	15 34	15 39	15 44	15 49	15 56	16 03	16 12
28	16 13	16 17	16 21	16 26	16 30	16 36	16 42	16 49	16 56	17 04	17 14	17 25	17 38	17 54
29	17 15	17 20	17 26	17 33	17 40	17 47	17 56	18 05	18 16	18 28	18 43	19 00	19 22	19 51
30	18 18	18 25	18 33	18 41	18 50	19 00	19 11	19 24	19 38	19 55	20 16	20 43	21 23	■
31	19 22	19 30	19 39	19 48	19 59	20 11	20 24	20 40	20 59	21 21	21 51	22 37	■	■
June 1	20 23	20 31	20 41	20 51	21 03	21 16	21 31	21 49	22 10	22 37	23 15	■	■	■
2	21 18	21 26	21 36	21 46	21 58	22 11	22 26	22 43	23 04	23 30	.. ..	■	■	■
3	22 05	22 13	22 22	22 31	22 42	22 53	23 07	23 22	23 39	.. ..	0 07	1 36	■	■
4	22 45	22 52	22 59	23 07	23 15	23 25	23 35	23 47	.. ..	0 01	0 28	1 06	■	■
5	23 18	23 23	23 29	23 35	23 41	23 48	23 56	.. ..	0 01	0 17	0 35	0 59	1 29	2 17
6	23 46	23 50	23 53	23 57	.. ..	.. ..	.. ..	0 05	0 15	0 26	0 38	0 53	1 11	1 33
7	.. ..	.. ..	.. ..	.. ..	0 02	0 07	0 12	0 18	0 24	0 31	0 39	0 48	0 59	1 12
8	0 11	0 13	0 15	0 17	0 20	0 22	0 25	0 28	0 31	0 35	0 39	0 44	0 49	0 55
9	0 35	0 35	0 35	0 36	0 36	0 36	0 37	0 37	0 38	0 38	0 39	0 39	0 40	0 41
10	0 59	0 58	0 56	0 54	0 53	0 51	0 49	0 47	0 44	0 41	0 38	0 35	0 31	0 27

MOONSET

Lat.	+40°	+42°	+44°	+46°	+48°	+50°	+52°	+54°	+56°	+58°	+60°	+62°	+64°	+66°
	h m	h m	h m	h m	h m	h m	h m	h m	h m	h m	h m	h m	h m	h m
May 17	20 35	20 44	20 53	21 03	21 14	21 26	21 40	21 57	22 16	22 40	23 12	.. ..	□	□
18	21 46	21 55	22 04	22 15	22 27	22 40	22 55	23 12	23 34	.. ..	.. ..	0 05	□	□
19	22 44	22 52	23 01	23 11	23 22	23 34	23 48	.. ..	.. ..	0 00	0 38	□	□	□
20	23 29	23 36	23 44	23 53	.. ..	.. ..	.. ..	0 04	0 23	0 46	1 16	2 02	□	□
21	.. ..	.. ..	.. ..	.. ..	0 02	0 12	0 23	0 36	0 51	1 08	1 29	1 56	2 33	3 56
22	0 05	0 10	0 16	0 23	0 30	0 38	0 47	0 57	1 07	1 20	1 34	1 51	2 12	2 39
23	0 33	0 37	0 42	0 46	0 52	0 57	1 03	1 10	1 18	1 26	1 36	1 47	2 00	2 15
24	0 56	0 59	1 02	1 05	1 08	1 12	1 16	1 20	1 25	1 30	1 36	1 42	1 50	1 59
25	1 17	1 19	1 20	1 21	1 23	1 24	1 26	1 28	1 30	1 32	1 35	1 38	1 41	1 45
26	1 37	1 37	1 36	1 36	1 36	1 36	1 35	1 35	1 35	1 34	1 34	1 33	1 33	1 32
27	1 56	1 55	1 53	1 51	1 49	1 47	1 45	1 42	1 39	1 36	1 33	1 29	1 25	1 19
28	2 17	2 14	2 11	2 07	2 03	1 59	1 55	1 50	1 45	1 39	1 32	1 24	1 16	1 05
29	2 40	2 35	2 30	2 25	2 20	2 14	2 07	2 00	1 52	1 42	1 32	1 20	1 05	0 48
30	3 07	3 01	2 54	2 47	2 40	2 32	2 23	2 13	2 01	1 48	1 33	1 15	0 52	0 23
31	3 39	3 32	3 24	3 15	3 06	2 56	2 44	2 31	2 16	1 58	1 37	1 09	0 29	■
June 1	4 19	4 10	4 01	3 51	3 41	3 28	3 15	2 59	2 40	2 17	1 47	1 00	■	■
2	5 07	4 59	4 49	4 38	4 27	4 13	3 58	3 41	3 19	2 52	2 14	■	■	■
3	6 05	5 56	5 47	5 37	5 25	5 12	4 57	4 40	4 20	3 54	3 18	1 48	■	■
4	7 10	7 02	6 54	6 45	6 34	6 23	6 10	5 56	5 38	5 18	4 51	4 13	■	■
5	8 19	8 13	8 06	7 58	7 50	7 41	7 31	7 20	7 07	6 52	6 34	6 11	5 41	4 55
6	9 30	9 25	9 20	9 15	9 09	9 02	8 55	8 47	8 38	8 28	8 16	8 02	7 45	7 24
7	10 40	10 37	10 34	10 31	10 27	10 23	10 19	10 14	10 09	10 03	9 56	9 48	9 39	9 28
8	11 51	11 50	11 49	11 47	11 46	11 44	11 43	11 41	11 39	11 36	11 34	11 31	11 27	11 23
9	13 02	13 03	13 04	13 04	13 05	13 06	13 07	13 08	13 09	13 10	13 12	13 13	13 15	13 17
10	14 16	14 18	14 21	14 24	14 27	14 30	14 34	14 38	14 42	14 47	14 53	14 59	15 07	15 16

□ indicates Moon continuously above horizon.
■ indicates Moon continuously below horizon.
.. .. indicates phenomenon will occur the next day.

UNIVERSAL TIME FOR MERIDIAN OF GREENWICH

MOONRISE

Lat.	−55°	−50°	−45°	−40°	−35°	−30°	−20°	−10°	0°	+10°	+20°	+30°	+35°	+40°
	h m	h m	h m	h m	h m	h m	h m	h m	h m	h m	h m	h m	h m	h m
June 8	.. ..	.. ..	.. ..	.. ..	.. ..	.. ..	.. ..	.. ..	.. ..	.. ..	.. ..	0 03	0 07	0 11
9	0 26	0 27	0 28	0 28	0 29	0 29	0 30	0 31	0 31	0 32	0 33	0 34	0 34	0 35
10	1 56	1 50	1 45	1 41	1 37	1 34	1 28	1 24	1 19	1 15	1 10	1 05	1 02	0 59
11	3 30	3 16	3 05	2 55	2 48	2 41	2 29	2 19	2 09	2 00	1 50	1 39	1 33	1 25
12	5 08	4 45	4 27	4 13	4 01	3 51	3 33	3 17	3 03	2 49	2 34	2 17	2 07	1 56
13	6 48	6 16	5 52	5 33	5 17	5 03	4 40	4 20	4 01	3 43	3 23	3 01	2 48	2 33
14	8 22	7 42	7 13	6 50	6 31	6 15	5 48	5 25	5 03	4 42	4 19	3 53	3 37	3 19
15	9 38	8 54	8 23	8 00	7 40	7 23	6 55	6 30	6 07	5 45	5 20	4 52	4 36	4 16
16	10 27	9 48	9 20	8 57	8 39	8 23	7 56	7 32	7 10	6 49	6 25	5 58	5 42	5 23
17	10 57	10 26	10 02	9 43	9 27	9 13	8 49	8 29	8 09	7 50	7 29	7 05	6 51	6 34
18	11 15	10 52	10 33	10 18	10 06	9 55	9 35	9 19	9 03	8 47	8 30	8 11	7 59	7 46
19	11 26	11 10	10 58	10 47	10 38	10 30	10 15	10 03	9 52	9 40	9 27	9 13	9 05	8 55
20	11 35	11 25	11 17	11 11	11 05	11 00	10 51	10 43	10 36	10 28	10 21	10 12	10 06	10 00
21	11 42	11 38	11 34	11 32	11 29	11 27	11 23	11 20	11 17	11 14	11 11	11 07	11 05	11 03
22	11 48	11 49	11 50	11 51	11 52	11 53	11 54	11 56	11 57	11 58	11 59	12 01	12 02	12 03
23	11 54	12 01	12 06	12 11	12 15	12 19	12 25	12 31	12 36	12 42	12 47	12 54	12 58	13 02
24	12 01	12 13	12 23	12 32	12 39	12 46	12 57	13 07	13 16	13 26	13 36	13 48	13 55	14 02
25	12 09	12 28	12 43	12 55	13 06	13 15	13 31	13 45	13 59	14 12	14 26	14 43	14 53	15 04
26	12 22	12 47	13 07	13 23	13 37	13 48	14 09	14 27	14 44	15 01	15 19	15 40	15 53	16 07
27	12 40	13 12	13 37	13 56	14 13	14 27	14 51	15 13	15 32	15 52	16 14	16 39	16 54	17 11
28	13 08	13 47	14 15	14 38	14 56	15 12	15 39	16 03	16 25	16 47	17 10	17 38	17 54	18 13
29	13 53	14 35	15 05	15 29	15 48	16 05	16 33	16 57	17 19	17 42	18 06	18 34	18 51	19 11
30	14 57	15 37	16 06	16 28	16 47	17 03	17 30	17 54	18 15	18 37	19 00	19 27	19 43	20 02
July 1	16 17	16 51	17 16	17 35	17 52	18 06	18 30	18 51	19 11	19 30	19 51	20 14	20 28	20 44
2	17 45	18 11	18 30	18 46	19 00	19 11	19 31	19 48	20 04	20 20	20 36	20 56	21 07	21 20

MOONSET

Lat.	−55°	−50°	−45°	−40°	−35°	−30°	−20°	−10°	0°	+10°	+20°	+30°	+35°	+40°
	h m	h m	h m	h m	h m	h m	h m	h m	h m	h m	h m	h m	h m	h m
June 8	12 32	12 28	12 24	12 22	12 19	12 17	12 13	12 10	12 07	12 03	12 00	11 56	11 54	11 51
9	12 39	12 42	12 44	12 45	12 47	12 48	12 50	12 52	12 54	12 56	12 58	13 00	13 01	13 02
10	12 47	12 57	13 04	13 10	13 15	13 20	13 28	13 36	13 43	13 50	13 57	14 05	14 10	14 16
11	12 58	13 14	13 27	13 38	13 47	13 56	14 10	14 22	14 34	14 46	14 59	15 14	15 22	15 32
12	13 12	13 36	13 56	14 11	14 25	14 36	14 56	15 14	15 30	15 47	16 04	16 25	16 37	16 51
13	13 34	14 07	14 32	14 52	15 09	15 24	15 49	16 10	16 30	16 51	17 12	17 37	17 52	18 10
14	14 10	14 51	15 21	15 44	16 03	16 20	16 48	17 11	17 34	17 56	18 20	18 48	19 05	19 24
15	15 09	15 52	16 22	16 46	17 06	17 23	17 51	18 15	18 38	19 00	19 24	19 52	20 09	20 28
16	16 28	17 07	17 34	17 56	18 14	18 30	18 56	19 18	19 39	20 00	20 22	20 47	21 02	21 20
17	17 58	18 28	18 51	19 09	19 24	19 37	19 59	20 18	20 36	20 53	21 12	21 33	21 46	22 00
18	19 28	19 49	20 06	20 20	20 31	20 41	20 58	21 13	21 27	21 41	21 55	22 12	22 21	22 32
19	20 52	21 06	21 18	21 27	21 35	21 42	21 53	22 04	22 13	22 23	22 33	22 44	22 51	22 58
20	22 12	22 20	22 26	22 31	22 35	22 39	22 45	22 51	22 56	23 01	23 06	23 13	23 16	23 20
21	23 28	23 30	23 31	23 32	23 33	23 33	23 34	23 35	23 36	23 37	23 38	23 39	23 40	23 40
22	.. ..	.. ..	.. ..	.. ..	.. ..	.. ..	.. ..	.. ..	.. ..	.. ..	.. ..	.. ..	.. ..	.. ..
23	0 43	0 38	0 35	0 32	0 29	0 27	0 23	0 19	0 16	0 12	0 09	0 05	0 03	0 00
24	1 57	1 47	1 38	1 31	1 25	1 20	1 11	1 03	0 55	0 48	0 40	0 31	0 26	0 20
25	3 13	2 56	2 43	2 32	2 23	2 14	2 00	1 48	1 36	1 25	1 13	0 59	0 51	0 42
26	4 32	4 08	3 49	3 34	3 21	3 10	2 52	2 35	2 20	2 05	1 49	1 30	1 20	1 07
27	5 51	5 20	4 56	4 38	4 22	4 08	3 45	3 25	3 07	2 48	2 29	2 06	1 53	1 38
28	7 09	6 30	6 03	5 41	5 23	5 07	4 41	4 18	3 57	3 36	3 14	2 48	2 32	2 15
29	8 17	7 35	7 05	6 41	6 22	6 05	5 37	5 13	4 51	4 28	4 04	3 36	3 20	3 00
30	9 09	8 28	7 59	7 36	7 17	7 00	6 32	6 09	5 46	5 24	5 00	4 32	4 15	3 56
July 1	9 43	9 09	8 43	8 22	8 05	7 50	7 25	7 03	6 42	6 21	5 59	5 33	5 18	5 00
2	10 06	9 39	9 18	9 01	8 47	8 34	8 13	7 54	7 36	7 19	6 59	6 37	6 24	6 09

.. .. indicates phenomenon will occur the next day.

UNIVERSAL TIME FOR MERIDIAN OF GREENWICH

MOONRISE

Lat.	+40°	+42°	+44°	+46°	+48°	+50°	+52°	+54°	+56°	+58°	+60°	+62°	+64°	+66°
	h m	h m	h m	h m	h m	h m	h m	h m	h m	h m	h m	h m	h m	h m
June 8	0 11	0 13	0 15	0 17	0 20	0 22	0 25	0 28	0 31	0 35	0 39	0 44	0 49	0 55
9	0 35	0 35	0 35	0 36	0 36	0 36	0 37	0 37	0 38	0 38	0 39	0 39	0 40	0 41
10	0 59	0 58	0 56	0 54	0 53	0 51	0 49	0 47	0 44	0 41	0 38	0 35	0 31	0 27
11	1 25	1 22	1 19	1 15	1 11	1 07	1 02	0 57	0 52	0 46	0 39	0 31	0 22	{00 11, 23 50}
12	1 56	1 51	1 45	1 40	1 33	1 27	1 19	1 11	1 02	0 52	0 40	00 26	{00 10, 23 53}	23 10
13	2 33	2 26	2 19	2 11	2 02	1 53	1 43	1 31	1 18	1 02	0 44	0 22	□	□
14	3 19	3 11	3 02	2 53	2 42	2 30	2 17	2 02	1 44	1 23	0 56	0 17	□	□
15	4 16	4 07	3 58	3 47	3 36	3 23	3 08	2 50	2 30	2 03	1 27	0 08	□	□
16	5 23	5 14	5 05	4 55	4 43	4 31	4 16	4 00	3 40	3 15	2 41	1 40	□	□
17	6 34	6 27	6 19	6 10	6 00	5 50	5 37	5 24	5 07	4 48	4 24	3 52	2 56	□
18	7 46	7 40	7 34	7 27	7 19	7 11	7 02	6 51	6 40	6 26	6 10	5 50	5 25	4 51
19	8 55	8 51	8 46	8 41	8 36	8 30	8 24	8 17	8 09	7 59	7 49	7 37	7 22	7 05
20	10 00	9 58	9 55	9 52	9 49	9 45	9 41	9 37	9 32	9 27	9 21	9 14	9 06	8 56
21	11 03	11 01	11 00	10 59	10 58	10 56	10 55	10 53	10 51	10 49	10 47	10 44	10 41	10 37
22	12 03	12 03	12 04	12 04	12 05	12 05	12 06	12 07	12 08	12 08	12 09	12 11	12 12	12 13
23	13 02	13 04	13 06	13 09	13 11	13 14	13 17	13 20	13 23	13 27	13 32	13 37	13 42	13 49
24	14 02	14 06	14 10	14 14	14 18	14 23	14 28	14 34	14 40	14 47	14 55	15 05	15 16	15 29
25	15 04	15 09	15 14	15 20	15 26	15 33	15 41	15 49	15 59	16 10	16 22	16 37	16 56	17 19
26	16 07	16 13	16 20	16 28	16 36	16 45	16 56	17 07	17 20	17 36	17 54	18 17	18 48	19 38
27	17 11	17 18	17 27	17 36	17 46	17 57	18 10	18 25	18 42	19 03	19 29	20 07	■	■
28	18 13	18 22	18 31	18 41	18 53	19 06	19 20	19 37	19 58	20 24	20 59	22 20	■	■
29	19 11	19 20	19 29	19 40	19 52	20 05	20 20	20 38	20 59	21 26	22 04	■	■	■
30	20 02	20 10	20 19	20 29	20 40	20 52	21 06	21 22	21 40	22 04	22 34	23 20	■	■
July 1	20 44	20 51	20 59	21 08	21 17	21 27	21 38	21 51	22 06	22 24	22 45	23 11	23 49	■
2	21 20	21 25	21 31	21 38	21 45	21 53	22 02	22 11	22 22	22 34	22 49	23 06	23 26	{01 19, 23 53}

MOONSET

Lat.	+40°	+42°	+44°	+46°	+48°	+50°	+52°	+54°	+56°	+58°	+60°	+62°	+64°	+66°
	h m	h m	h m	h m	h m	h m	h m	h m	h m	h m	h m	h m	h m	h m
June 8	11 51	11 50	11 49	11 47	11 46	11 44	11 43	11 41	11 39	11 36	11 34	11 31	11 27	11 23
9	13 02	13 03	13 04	13 04	13 05	13 06	13 07	13 08	13 09	13 10	13 12	13 13	13 15	13 17
10	14 16	14 18	14 21	14 24	14 27	14 30	14 34	14 38	14 42	14 47	14 53	14 59	15 07	15 16
11	15 32	15 36	15 41	15 46	15 52	15 58	16 04	16 11	16 20	16 29	16 40	16 52	17 07	17 26
12	16 51	16 57	17 04	17 11	17 19	17 28	17 38	17 49	18 01	18 16	18 33	18 55	19 23	20 04
13	18 10	18 17	18 26	18 35	18 45	18 57	19 10	19 24	19 42	20 03	20 29	21 07	□	□
14	19 24	19 33	19 42	19 52	20 04	20 17	20 32	20 49	21 10	21 36	22 12	23 31	□	□
15	20 28	20 37	20 46	20 56	21 08	21 21	21 35	21 52	22 12	22 37	23 12	.. ..	□	□
16	21 20	21 27	21 36	21 45	21 55	22 06	22 19	22 33	22 50	23 09	23 34	0 13	□	□
17	22 00	22 07	22 13	22 21	22 29	22 38	22 47	22 58	23 11	23 25	23 42	0 07	1 03	□
18	22 32	22 37	22 42	22 47	22 53	23 00	23 07	23 15	23 24	23 34	23 45	{00 03, 23 58}	0 28	1 04
19	22 58	23 01	23 05	23 08	23 12	23 17	23 21	23 27	23 32	23 39	23 46	23 54	0 14	0 33
20	23 20	23 22	23 24	23 26	23 28	23 30	23 33	23 35	23 38	23 42	23 45	23 50	{00 03, 23 54}	0 14
21	23 40	23 41	23 41	23 41	23 42	23 42	23 42	23 43	23 43	23 44	23 45	23 45	23 46	{00 00, 23 47}
22	.. ..	23 59	23 58	23 56	23 55	23 54	23 52	23 50	23 48	23 46	23 44	23 41	23 38	23 34
23	0 00	.. ..	.. ..	.. ..	.. ..	.. ..	.. ..	23 58	23 53	23 48	23 43	23 36	23 29	23 21
24	0 20	0 18	0 15	0 12	0 09	0 05	0 02	.. ..	23 59	23 51	23 42	23 32	23 20	23 05
25	0 42	0 38	0 34	0 29	0 24	0 19	0 13	0 07	.. ..	23 56	23 43	23 27	23 08	22 44
26	1 07	1 02	0 56	0 50	0 43	0 36	0 27	0 18	0 08	.. ..	23 45	23 22	22 50	21 59
27	1 38	1 31	1 23	1 15	1 07	0 57	0 46	0 34	0 21	0 05	23 53	23 15	■	■
28	2 15	2 07	1 58	1 48	1 38	1 26	1 13	0 58	0 41	0 20	.. ..	22 52	■	■
29	3 00	2 52	2 42	2 32	2 20	2 07	1 52	1 35	1 14	0 48	0 12	■	■	■
30	3 56	3 47	3 37	3 27	3 15	3 02	2 47	2 29	2 08	1 41	1 03	■	■	■
July 1	5 00	4 52	4 43	4 33	4 22	4 10	3 57	3 41	3 23	3 00	2 30	1 44	■	■
2	6 09	6 02	5 55	5 47	5 38	5 28	5 17	5 05	4 51	4 34	4 13	3 47	3 10	1 41

□ indicates Moon continuously above horizon.
■ indicates Moon continuously below horizon.
.. .. indicates phenomenon will occur the next day.

UNIVERSAL TIME FOR MERIDIAN OF GREENWICH

MOONRISE

Lat.	−55°	−50°	−45°	−40°	−35°	−30°	−20°	−10°	0°	+10°	+20°	+30°	+35°	+40°
	h m	h m	h m	h m	h m	h m	h m	h m	h m	h m	h m	h m	h m	h m
July 1	16 17	16 51	17 16	17 35	17 52	18 06	18 30	18 51	19 11	19 30	19 51	20 14	20 28	20 44
2	17 45	18 11	18 30	18 46	19 00	19 11	19 31	19 48	20 04	20 20	20 36	20 56	21 07	21 20
3	19 16	19 33	19 47	19 58	20 07	20 16	20 30	20 43	20 54	21 06	21 18	21 32	21 40	21 49
4	20 45	20 55	21 03	21 09	21 15	21 20	21 28	21 36	21 43	21 50	21 57	22 05	22 10	22 15
5	22 13	22 16	22 18	22 20	22 22	22 23	22 25	22 28	22 30	22 32	22 34	22 36	22 38	22 39
6	23 42	23 37	23 34	23 31	23 29	23 26	23 23	23 20	23 17	23 14	23 11	23 07	23 05	23 03
7	.. ..	.. ..	.. ..	.. ..	.. ..	.. ..	.. ..	.. ..	.. ..	23 57	23 49	23 39	23 34	23 28
8	1 12	1 01	0 51	0 44	0 37	0 31	0 21	0 13	0 05	.. ..	.. ..	.. ..	.. ..	23 56
9	2 46	2 27	2 11	1 58	1 48	1 38	1 23	1 09	0 56	0 43	0 30	0 15	0 06	.. ..
10	4 23	3 54	3 33	3 15	3 01	2 48	2 27	2 08	1 51	1 34	1 16	0 55	0 43	0 29
11	5 58	5 21	4 53	4 32	4 14	3 58	3 33	3 10	2 50	2 29	2 08	1 42	1 28	1 11
12	7 21	6 37	6 07	5 43	5 23	5 07	4 39	4 14	3 52	3 29	3 05	2 38	2 21	2 02
13	8 20	7 38	7 08	6 45	6 26	6 09	5 41	5 17	4 54	4 32	4 08	3 40	3 23	3 04
14	8 57	8 22	7 56	7 35	7 18	7 03	6 37	6 15	5 55	5 34	5 12	4 46	4 31	4 13
15	9 19	8 52	8 31	8 15	8 00	7 48	7 27	7 08	6 51	6 33	6 15	5 53	5 40	5 25
16	9 33	9 14	8 58	8 46	8 35	8 26	8 10	7 55	7 42	7 28	7 14	6 57	6 47	6 36
17	9 42	9 30	9 20	9 12	9 05	8 58	8 47	8 37	8 28	8 19	8 09	7 58	7 52	7 44
18	9 50	9 43	9 38	9 34	9 30	9 27	9 21	9 16	9 11	9 07	9 01	8 56	8 52	8 48
19	9 56	9 55	9 55	9 54	9 54	9 54	9 53	9 53	9 52	9 52	9 51	9 51	9 50	9 50
20	10 02	10 07	10 11	10 14	10 17	10 20	10 24	10 28	10 32	10 36	10 40	10 45	10 47	10 51
21	10 08	10 19	10 27	10 34	10 41	10 46	10 56	11 04	11 12	11 20	11 29	11 38	11 44	11 51
22	10 16	10 33	10 46	10 57	11 06	11 14	11 29	11 41	11 53	12 05	12 18	12 33	12 42	12 52
23	10 27	10 50	11 08	11 22	11 35	11 46	12 05	12 22	12 37	12 53	13 10	13 29	13 41	13 54
24	10 42	11 12	11 35	11 53	12 09	12 22	12 45	13 05	13 24	13 43	14 03	14 27	14 41	14 57
25	11 05	11 42	12 09	12 31	12 49	13 04	13 31	13 53	14 15	14 36	14 59	15 26	15 42	16 00

MOONSET

Lat.	−55°	−50°	−45°	−40°	−35°	−30°	−20°	−10°	0°	+10°	+20°	+30°	+35°	+40°
	h m	h m	h m	h m	h m	h m	h m	h m	h m	h m	h m	h m	h m	h m
July 1	9 43	9 09	8 43	8 22	8 05	7 50	7 25	7 03	6 42	6 21	5 59	5 33	5 18	5 00
2	10 06	9 39	9 18	9 01	8 47	8 34	8 13	7 54	7 36	7 19	6 59	6 37	6 24	6 09
3	10 21	10 02	9 46	9 34	9 23	9 13	8 56	8 42	8 28	8 14	7 59	7 42	7 32	7 20
4	10 32	10 19	10 10	10 01	9 54	9 48	9 37	9 27	9 18	9 08	8 58	8 47	8 40	8 32
5	10 40	10 35	10 30	10 26	10 23	10 20	10 14	10 10	10 05	10 00	9 56	9 50	9 47	9 43
6	10 48	10 49	10 49	10 50	10 50	10 50	10 51	10 51	10 52	10 52	10 53	10 53	10 53	10 54
7	10 56	11 03	11 09	11 14	11 18	11 22	11 28	11 34	11 39	11 45	11 51	11 57	12 01	12 05
8	11 05	11 19	11 30	11 40	11 48	11 55	12 07	12 18	12 29	12 39	12 50	13 03	13 10	13 19
9	11 17	11 39	11 56	12 10	12 22	12 33	12 51	13 07	13 22	13 37	13 53	14 11	14 22	14 35
10	11 35	12 05	12 28	12 47	13 02	13 16	13 39	14 00	14 18	14 37	14 58	15 22	15 35	15 52
11	12 04	12 43	13 11	13 33	13 51	14 07	14 34	14 57	15 19	15 41	16 04	16 31	16 47	17 06
12	12 52	13 35	14 06	14 30	14 49	15 06	15 35	15 59	16 22	16 44	17 09	17 37	17 54	18 13
13	14 02	14 43	15 13	15 36	15 55	16 11	16 38	17 02	17 24	17 45	18 09	18 35	18 51	19 10
14	15 29	16 03	16 28	16 47	17 04	17 18	17 42	18 03	18 22	18 41	19 02	19 25	19 39	19 54
15	16 59	17 25	17 44	18 00	18 13	18 24	18 44	19 00	19 16	19 31	19 48	20 06	20 17	20 30
16	18 27	18 45	18 58	19 09	19 19	19 27	19 41	19 53	20 05	20 16	20 28	20 41	20 49	20 58
17	19 50	20 01	20 09	20 15	20 21	20 26	20 35	20 42	20 49	20 56	21 04	21 12	21 17	21 22
18	21 09	21 13	21 16	21 18	21 21	21 22	21 26	21 28	21 31	21 34	21 36	21 39	21 41	21 43
19	22 25	22 23	22 21	22 19	22 18	22 17	22 15	22 13	22 11	22 10	22 08	22 06	22 04	22 03
20	23 40	23 32	23 25	23 20	23 15	23 11	23 03	22 57	22 51	22 45	22 39	22 32	22 28	22 23
21	.. ..	.. ..	.. ..	.. ..	.. ..	.. ..	23 52	23 42	23 32	23 22	23 11	22 59	22 52	22 44
22	0 56	0 41	0 30	0 20	0 12	0 05	.. ..	.. ..	.. ..	.. ..	23 46	23 29	23 19	23 08
23	2 13	1 52	1 35	1 22	1 10	1 00	0 43	0 28	0 14	0 00	.. ..	.. ..	23 50	23 36
24	3 32	3 04	2 42	2 24	2 10	1 57	1 35	1 17	0 59	0 42	0 24	0 02	.. ..	.. ..
25	4 51	4 15	3 48	3 28	3 10	2 55	2 30	2 08	1 48	1 28	1 06	0 41	0 27	0 10

.. .. indicates phenomenon will occur the next day.

UNIVERSAL TIME FOR MERIDIAN OF GREENWICH

MOONRISE

Lat.	+40°	+42°	+44°	+46°	+48°	+50°	+52°	+54°	+56°	+58°	+60°	+62°	+64°	+66°
	h m	h m	h m	h m	h m	h m	h m	h m	h m	h m	h m	h m	h m	h m
July 1	20 44	20 51	20 59	21 08	21 17	21 27	21 38	21 51	22 06	22 24	22 45	23 11	23 49	■
2	21 20	21 25	21 31	21 38	21 45	21 53	22 02	22 11	22 22	22 34	22 49	23 06	23 26	{01 19 / 23 53}
3	21 49	21 53	21 58	22 02	22 07	22 13	22 19	22 25	22 33	22 41	22 50	23 01	23 13	23 28
4	22 15	22 18	22 20	22 23	22 26	22 29	22 33	22 36	22 40	22 45	22 50	22 56	23 03	23 10
5	22 39	22 40	22 41	22 42	22 43	22 43	22 45	22 46	22 47	22 48	22 50	22 51	22 53	22 56
6	23 03	23 02	23 01	23 00	22 59	22 58	22 56	22 55	22 53	22 51	22 49	22 47	22 44	22 41
7	23 28	23 25	23 22	23 19	23 16	23 13	23 09	23 05	23 00	22 55	22 49	22 43	22 35	22 26
8	23 56	23 52	23 47	23 42	23 36	23 30	23 24	23 17	23 09	23 00	22 50	22 38	22 24	22 08
9	.. ..	.. ..	.. ..	.. ..	.. ..	23 53	23 44	23 34	23 22	23 08	22 52	22 34	22 10	21 39
10	0 29	0 23	0 17	0 09	0 02	.. ..	.. ..	23 59	23 43	23 23	23 00	22 29	21 40	□
11	1 11	1 03	0 55	0 46	0 36	0 25	0 13	.. ..	.. ..	23 53	23 20	22 22	□	□
12	2 02	1 53	1 44	1 34	1 22	1 10	0 55	0 38	0 18	.. ..	.. ..	□	□	□
13	3 04	2 55	2 45	2 35	2 23	2 10	1 55	1 38	1 17	0 51	0 14	□	□	□
14	4 13	4 05	3 56	3 47	3 36	3 25	3 11	2 56	2 38	2 16	1 48	1 05	□	□
15	5 25	5 19	5 12	5 04	4 55	4 46	4 35	4 23	4 10	3 54	3 34	3 10	2 37	1 39
16	6 36	6 31	6 26	6 20	6 14	6 07	5 59	5 51	5 41	5 30	5 18	5 03	4 44	4 21
17	7 44	7 41	7 37	7 33	7 29	7 25	7 20	7 14	7 08	7 01	6 54	6 45	6 34	6 22
18	8 48	8 47	8 45	8 43	8 41	8 39	8 36	8 33	8 30	8 27	8 23	8 19	8 14	8 08
19	9 50	9 50	9 50	9 50	9 50	9 49	9 49	9 49	9 49	9 48	9 48	9 48	9 47	9 47
20	10 51	10 52	10 53	10 55	10 57	10 59	11 01	11 03	11 05	11 08	11 11	11 15	11 19	11 24
21	11 51	11 54	11 57	12 00	12 04	12 08	12 12	12 17	12 22	12 28	12 35	12 42	12 51	13 02
22	12 52	12 56	13 01	13 06	13 12	13 18	13 24	13 32	13 40	13 50	14 00	14 13	14 29	14 47
23	13 54	14 00	14 06	14 13	14 21	14 29	14 38	14 49	15 01	15 14	15 30	15 50	16 15	16 51
24	14 57	15 05	15 12	15 21	15 31	15 41	15 53	16 07	16 22	16 41	17 04	17 36	18 26	■
25	16 00	16 08	16 18	16 28	16 39	16 51	17 05	17 22	17 41	18 05	18 38	19 35	■	■

MOONSET

Lat.	+40°	+42°	+44°	+46°	+48°	+50°	+52°	+54°	+56°	+58°	+60°	+62°	+64°	+66°
	h m	h m	h m	h m	h m	h m	h m	h m	h m	h m	h m	h m	h m	h m
July 1	5 00	4 52	4 43	4 33	4 22	4 10	3 57	3 41	3 23	3 00	2 30	1 44	■	■
2	6 09	6 02	5 55	5 47	5 38	5 28	5 17	5 05	4 51	4 34	4 13	3 47	3 10	1 41
3	7 20	7 15	7 10	7 04	6 57	6 50	6 42	6 33	6 23	6 11	5 58	5 42	5 22	4 57
4	8 32	8 29	8 25	8 21	8 17	8 12	8 07	8 01	7 55	7 48	7 40	7 30	7 19	7 06
5	9 43	9 41	9 40	9 38	9 36	9 33	9 31	9 28	9 25	9 22	9 18	9 14	9 09	9 03
6	10 54	10 54	10 54	10 54	10 54	10 55	10 55	10 55	10 55	10 55	10 56	10 56	10 56	10 57
7	12 05	12 07	12 09	12 12	12 14	12 17	12 19	12 23	12 26	12 30	12 34	12 39	12 45	12 52
8	13 19	13 23	13 27	13 31	13 36	13 41	13 47	13 53	14 00	14 08	14 17	14 27	14 40	14 54
9	14 35	14 40	14 46	14 53	15 00	15 08	15 17	15 26	15 37	15 50	16 05	16 23	16 46	17 16
10	15 52	15 59	16 07	16 15	16 25	16 35	16 47	17 01	17 16	17 35	17 58	18 28	19 17	□
11	17 06	17 15	17 24	17 34	17 45	17 58	18 12	18 29	18 48	19 13	19 46	20 44	□	□
12	18 13	18 22	18 32	18 42	18 54	19 07	19 22	19 40	20 01	20 27	21 04	□	□	□
13	19 10	19 18	19 27	19 37	19 47	19 59	20 13	20 29	20 47	21 09	21 38	22 21	□	□
14	19 54	20 01	20 09	20 17	20 26	20 36	20 47	21 00	21 14	21 30	21 50	22 16	22 50	23 48
15	20 30	20 35	20 41	20 47	20 54	21 02	21 10	21 19	21 30	21 41	21 55	22 11	22 30	22 54
16	20 58	21 02	21 06	21 11	21 15	21 21	21 26	21 33	21 40	21 47	21 56	22 06	22 18	22 32
17	21 22	21 24	21 27	21 30	21 32	21 36	21 39	21 43	21 47	21 51	21 56	22 02	22 08	22 16
18	21 43	21 44	21 45	21 46	21 47	21 48	21 49	21 51	21 52	21 54	21 56	21 58	22 00	22 03
19	22 03	22 03	22 02	22 01	22 01	22 00	21 59	21 58	21 57	21 56	21 55	21 53	21 52	21 50
20	22 23	22 21	22 19	22 17	22 14	22 11	22 09	22 05	22 02	21 58	21 54	21 49	21 43	21 37
21	22 44	22 41	22 37	22 33	22 29	22 24	22 19	22 14	22 07	22 01	21 53	21 44	21 34	21 22
22	23 08	23 03	22 58	22 52	22 46	22 39	22 32	22 24	22 15	22 05	21 53	21 39	21 23	21 03
23	23 36	23 30	23 23	23 15	23 07	22 59	22 49	22 38	22 25	22 11	21 54	21 34	21 08	20 32
24	.. ..	.. ..	23 54	23 45	23 35	23 24	23 12	22 58	22 42	22 23	21 59	21 27	20 36	■
25	0 10	0 02	.. ..	.. ..	.. ..	.. ..	23 45	23 29	23 09	22 44	22 11	21 14	■	■

□ indicates Moon continuously above horizon.
■ indicates Moon continuously below horizon.
.. .. indicates phenomenon will occur the next day.

UNIVERSAL TIME FOR MERIDIAN OF GREENWICH

MOONRISE

Lat.	−55°	−50°	−45°	−40°	−35°	−30°	−20°	−10°	0°	+10°	+20°	+30°	+35°	+40°
	h m	h m	h m	h m	h m	h m	h m	h m	h m	h m	h m	h m	h m	h m
July 24	10 42	11 12	11 35	11 53	12 09	12 22	12 45	13 05	13 24	13 43	14 03	14 27	14 41	14 57
25	11 05	11 42	12 09	12 31	12 49	13 04	13 31	13 53	14 15	14 36	14 59	15 26	15 42	16 00
26	11 43	12 24	12 54	13 18	13 37	13 54	14 22	14 46	15 08	15 31	15 55	16 23	16 40	17 00
27	12 39	13 21	13 51	14 14	14 33	14 50	15 18	15 42	16 04	16 26	16 50	17 18	17 35	17 54
28	13 54	14 31	14 58	15 19	15 37	15 52	16 18	16 40	17 00	17 21	17 43	18 08	18 23	18 40
29	15 22	15 51	16 13	16 30	16 45	16 57	17 19	17 38	17 55	18 12	18 31	18 52	19 04	19 19
30	16 54	17 14	17 30	17 43	17 54	18 04	18 20	18 34	18 48	19 01	19 15	19 31	19 40	19 51
31	18 26	18 38	18 48	18 56	19 03	19 10	19 20	19 29	19 38	19 46	19 56	20 06	20 12	20 18
Aug. 1	19 57	20 02	20 06	20 09	20 12	20 15	20 19	20 23	20 26	20 30	20 34	20 38	20 41	20 43
2	21 27	21 25	21 23	21 21	21 20	21 19	21 17	21 16	21 14	21 13	21 11	21 10	21 09	21 07
3	22 58	22 48	22 41	22 34	22 29	22 24	22 16	22 09	22 03	21 56	21 49	21 41	21 37	21 32
4	.. ..	.. ..	.. ..	23 49	23 39	23 31	23 17	23 04	22 53	22 41	22 29	22 16	22 08	21 59
5	0 31	0 14	0 00	.. ..	.. ..	.. ..	.. ..	.. ..	23 46	23 30	23 13	22 54	22 43	22 30
6	2 07	1 41	1 21	1 05	0 51	0 39	0 19	0 02	.. ..	.. ..	.. ..	23 39	23 24	23 08
7	3 42	3 07	2 41	2 21	2 04	1 49	1 24	1 03	0 43	0 23	0 03	.. ..	.. ..	23 55
8	5 08	4 26	3 56	3 33	3 13	2 57	2 29	2 05	1 43	1 21	0 57	0 30	0 14	.. ..
9	6 15	5 31	5 00	4 37	4 17	4 00	3 31	3 07	2 44	2 22	1 57	1 29	1 12	0 52
10	6 58	6 20	5 52	5 30	5 11	4 56	4 29	4 06	3 44	3 23	3 00	2 33	2 17	1 58
11	7 24	6 53	6 31	6 12	5 57	5 43	5 20	5 00	4 41	4 22	4 02	3 38	3 25	3 08
12	7 40	7 18	7 00	6 46	6 34	6 23	6 05	5 48	5 33	5 18	5 02	4 43	4 32	4 19
13	7 50	7 35	7 23	7 13	7 05	6 57	6 44	6 32	6 21	6 10	5 58	5 45	5 37	5 28
14	7 58	7 50	7 43	7 37	7 32	7 27	7 19	7 12	7 06	6 59	6 52	6 44	6 39	6 34
15	8 05	8 02	8 00	7 58	7 56	7 54	7 52	7 49	7 47	7 45	7 43	7 40	7 38	7 37
16	8 11	8 14	8 16	8 18	8 19	8 21	8 23	8 25	8 28	8 30	8 32	8 35	8 36	8 38
17	8 17	8 25	8 32	8 38	8 43	8 47	8 55	9 01	9 08	9 14	9 21	9 29	9 33	9 38

MOONSET

Lat.	−55°	−50°	−45°	−40°	−35°	−30°	−20°	−10°	0°	+10°	+20°	+30°	+35°	+40°
	h m	h m	h m	h m	h m	h m	h m	h m	h m	h m	h m	h m	h m	h m
July 24	3 32	3 04	2 42	2 24	2 10	1 57	1 35	1 17	0 59	0 42	0 24	0 02	.. ..	.. ..
25	4 51	4 15	3 48	3 28	3 10	2 55	2 30	2 08	1 48	1 28	1 06	0 41	0 27	0 10
26	6 03	5 22	4 52	4 29	4 10	3 54	3 26	3 02	2 40	2 18	1 54	1 27	1 11	0 52
27	7 02	6 20	5 50	5 26	5 07	4 50	4 22	3 58	3 35	3 12	2 48	2 20	2 03	1 43
28	7 43	7 06	6 38	6 16	5 58	5 42	5 16	4 53	4 31	4 10	3 46	3 19	3 03	2 44
29	8 10	7 40	7 17	6 58	6 43	6 29	6 06	5 46	5 27	5 08	4 47	4 23	4 09	3 53
30	8 28	8 05	7 48	7 34	7 21	7 11	6 52	6 36	6 21	6 05	5 48	5 29	5 18	5 05
31	8 40	8 25	8 13	8 03	7 55	7 47	7 34	7 23	7 12	7 01	6 49	6 35	6 27	6 18
Aug. 1	8 49	8 41	8 35	8 30	8 25	8 21	8 14	8 07	8 01	7 55	7 48	7 41	7 36	7 31
2	8 57	8 56	8 55	8 54	8 53	8 52	8 51	8 50	8 49	8 48	8 47	8 45	8 44	8 44
3	9 05	9 10	9 14	9 18	9 21	9 24	9 29	9 33	9 37	9 41	9 45	9 50	9 53	9 56
4	9 13	9 25	9 35	9 43	9 50	9 57	10 08	10 17	10 26	10 35	10 45	10 56	11 02	11 10
5	9 24	9 44	9 59	10 12	10 23	10 33	10 49	11 04	11 18	11 32	11 46	12 03	12 13	12 25
6	9 40	10 08	10 29	10 46	11 01	11 14	11 36	11 55	12 13	12 31	12 50	13 13	13 26	13 41
7	10 04	10 40	11 07	11 29	11 46	12 02	12 28	12 50	13 11	13 32	13 55	14 22	14 37	14 55
8	10 44	11 27	11 57	12 21	12 40	12 57	13 25	13 50	14 12	14 35	14 59	15 28	15 44	16 04
9	11 45	12 28	12 59	13 22	13 42	13 59	14 27	14 51	15 13	15 36	16 00	16 27	16 44	17 03
10	13 06	13 43	14 10	14 31	14 49	15 04	15 30	15 52	16 12	16 32	16 54	17 19	17 34	17 50
11	14 35	15 03	15 25	15 42	15 57	16 10	16 31	16 50	17 07	17 24	17 42	18 03	18 15	18 28
12	16 03	16 24	16 40	16 53	17 04	17 13	17 29	17 44	17 57	18 10	18 24	18 39	18 49	18 59
13	17 28	17 41	17 51	18 00	18 07	18 14	18 25	18 34	18 43	18 52	19 01	19 11	19 17	19 24
14	18 49	18 55	19 00	19 04	19 08	19 11	19 17	19 21	19 26	19 30	19 35	19 40	19 43	19 46
15	20 06	20 06	20 06	20 06	20 06	20 07	20 07	20 07	20 07	20 07	20 07	20 07	20 07	20 07
16	21 22	21 16	21 11	21 07	21 04	21 01	20 56	20 51	20 47	20 42	20 38	20 33	20 30	20 27
17	22 38	22 25	22 16	22 08	22 01	21 55	21 45	21 35	21 27	21 19	21 10	21 00	20 54	20 47

.. .. indicates phenomenon will occur the next day.

UNIVERSAL TIME FOR MERIDIAN OF GREENWICH

MOONRISE

Lat.	+40°	+42°	+44°	+46°	+48°	+50°	+52°	+54°	+56°	+58°	+60°	+62°	+64°	+66°
	h m	h m	h m	h m	h m	h m	h m	h m	h m	h m	h m	h m	h m	h m
July 24	14 57	15 05	15 12	15 21	15 31	15 41	15 53	16 07	16 22	16 41	17 04	17 36	18 26	■
25	16 00	16 08	16 18	16 28	16 39	16 51	17 05	17 22	17 41	18 05	18 38	19 35	■	■
26	17 00	17 09	17 18	17 29	17 41	17 54	18 10	18 27	18 49	19 16	19 56	■	■	■
27	17 54	18 02	18 12	18 22	18 34	18 46	19 01	19 18	19 38	20 04	20 38	21 43	■	■
28	18 40	18 48	18 56	19 05	19 15	19 26	19 39	19 53	20 10	20 29	20 54	21 27	22 23	■
29	19 19	19 25	19 32	19 39	19 47	19 56	20 05	20 16	20 29	20 43	21 00	21 20	21 45	22 20
30	19 51	19 55	20 00	20 06	20 12	20 18	20 25	20 32	20 41	20 51	21 01	21 14	21 29	21 47
31	20 18	20 21	20 25	20 28	20 32	20 36	20 40	20 44	20 50	20 55	21 02	21 09	21 18	21 27
Aug. 1	20 43	20 45	20 46	20 47	20 49	20 51	20 52	20 54	20 56	20 59	21 01	21 04	21 08	21 12
2	21 07	21 07	21 07	21 06	21 05	21 05	21 04	21 03	21 03	21 02	21 01	21 00	20 59	20 57
3	21 32	21 30	21 28	21 25	21 22	21 20	21 16	21 13	21 09	21 05	21 00	20 55	20 49	20 42
4	21 59	21 55	21 51	21 46	21 42	21 36	21 31	21 24	21 17	21 09	21 01	20 51	20 39	20 25
5	22 30	22 25	22 19	22 12	22 05	21 57	21 49	21 39	21 28	21 16	21 02	20 46	20 26	20 00
6	23 08	23 01	22 53	22 45	22 35	22 25	22 14	22 01	21 46	21 28	21 07	20 40	20 03	18 34
7	23 55	23 47	23 38	23 28	23 17	23 04	22 50	22 34	22 15	21 52	21 21	20 33	□	□
8	.. ..	.. ..	.. ..	.. ..	.. ..	23 58	23 43	23 25	23 04	22 37	21 59	□	□	□
9	0 52	0 44	0 34	0 23	0 11	.. ..	.. ..	.. ..	.. ..	23 52	23 20	22 24	□	□
10	1 58	1 50	1 40	1 30	1 19	1 07	0 53	0 36	0 17	.. ..	.. ..	.. ..	23 46	□
11	3 08	3 01	2 53	2 45	2 35	2 25	2 13	2 00	1 45	1 26	1 03	0 33	.. ..	□
12	4 19	4 14	4 08	4 01	3 54	3 46	3 37	3 27	3 16	3 03	2 48	2 30	2 07	1 36
13	5 28	5 24	5 20	5 15	5 10	5 05	4 59	4 52	4 44	4 36	4 26	4 15	4 02	3 46
14	6 34	6 31	6 29	6 26	6 23	6 20	6 17	6 13	6 08	6 04	5 58	5 52	5 45	5 37
15	7 37	7 36	7 35	7 34	7 33	7 32	7 31	7 30	7 29	7 27	7 25	7 23	7 21	7 19
16	8 38	8 39	8 40	8 40	8 41	8 42	8 44	8 45	8 46	8 48	8 50	8 52	8 54	8 56
17	9 38	9 41	9 43	9 46	9 49	9 52	9 55	9 59	10 03	10 08	10 13	10 19	10 26	10 34

MOONSET

Lat.	+40°	+42°	+44°	+46°	+48°	+50°	+52°	+54°	+56°	+58°	+60°	+62°	+64°	+66°
	h m	h m	h m	h m	h m	h m	h m	h m	h m	h m	h m	h m	h m	h m
July 24	.. ..	.. ..	23 54	23 45	23 35	23 24	23 12	22 58	22 42	22 23	21 59	21 27	20 36	■
25	0 10	0 02	.. ..	.. ..	.. ..	.. ..	23 45	23 29	23 09	22 44	22 11	21 14	■	■
26	0 52	0 43	0 34	0 24	0 12	0 00	.. ..	.. ..	23 54	23 26	22 47	■	■	■
27	1 43	1 34	1 25	1 14	1 02	0 49	0 33	0 16	.. ..	.. ..	.. ..	22 57	■	■
28	2 44	2 36	2 26	2 16	2 05	1 52	1 38	1 21	1 01	0 36	0 02	.. ..	■	■
29	3 53	3 45	3 37	3 28	3 19	3 08	2 56	2 42	2 26	2 07	1 43	1 10	0 15	■
30	5 05	4 59	4 53	4 46	4 38	4 30	4 21	4 11	3 59	3 46	3 30	3 11	2 46	2 12
31	6 18	6 14	6 10	6 05	6 00	5 54	5 48	5 41	5 34	5 25	5 15	5 04	4 50	4 33
Aug. 1	7 31	7 29	7 27	7 24	7 21	7 18	7 15	7 11	7 07	7 03	6 57	6 52	6 45	6 37
2	8 44	8 43	8 43	8 42	8 42	8 41	8 41	8 40	8 39	8 38	8 37	8 36	8 35	8 34
3	9 56	9 57	9 59	10 01	10 02	10 04	10 06	10 09	10 11	10 14	10 17	10 21	10 25	10 30
4	11 10	11 13	11 16	11 20	11 24	11 29	11 34	11 39	11 45	11 52	11 59	12 08	12 18	12 31
5	12 25	12 30	12 35	12 41	12 48	12 55	13 03	13 12	13 21	13 33	13 46	14 01	14 21	14 45
6	13 41	13 48	13 55	14 03	14 12	14 22	14 33	14 45	15 00	15 16	15 37	16 03	16 40	18 08
7	14 55	15 04	15 12	15 22	15 33	15 45	15 59	16 15	16 33	16 57	17 27	18 15	□	□
8	16 04	16 13	16 22	16 33	16 45	16 58	17 13	17 31	17 52	18 19	18 57	□	□	□
9	17 03	17 11	17 21	17 31	17 42	17 55	18 09	18 26	18 46	19 10	19 43	20 39	□	□
10	17 50	17 58	18 06	18 15	18 25	18 36	18 48	19 02	19 18	19 37	20 00	20 31	21 18	□
11	18 28	18 35	18 41	18 48	18 56	19 05	19 14	19 24	19 36	19 50	20 06	20 25	20 49	21 21
12	18 59	19 03	19 08	19 14	19 19	19 25	19 32	19 40	19 48	19 57	20 08	20 20	20 35	20 52
13	19 24	19 27	19 30	19 34	19 37	19 41	19 46	19 51	19 56	20 02	20 08	20 16	20 24	20 34
14	19 46	19 48	19 49	19 51	19 53	19 55	19 57	19 59	20 02	20 04	20 08	20 11	20 15	20 20
15	20 07	20 07	20 07	20 07	20 07	20 07	20 07	20 07	20 07	20 07	20 07	20 06	20 06	20 06
16	20 27	20 25	20 24	20 22	20 20	20 18	20 16	20 14	20 11	20 08	20 05	20 02	19 58	19 53
17	20 47	20 44	20 41	20 38	20 34	20 30	20 26	20 22	20 16	20 11	20 04	19 57	19 49	19 39

□ indicates Moon continuously above horizon.
■ indicates Moon continuously below horizon.
.. .. indicates phenomenon will occur the next day.

UNIVERSAL TIME FOR MERIDIAN OF GREENWICH

MOONRISE

Lat.	−55°	−50°	−45°	−40°	−35°	−30°	−20°	−10°	0°	+10°	+20°	+30°	+35°	+40°
	h m	h m	h m	h m	h m	h m	h m	h m	h m	h m	h m	h m	h m	h m
Aug. 16	8 11	8 14	8 16	8 18	8 19	8 21	8 23	8 25	8 28	8 30	8 32	8 35	8 36	8 38
17	8 17	8 25	8 32	8 38	8 43	8 47	8 55	9 01	9 08	9 14	9 21	9 29	9 33	9 38
18	8 24	8 38	8 50	8 59	9 07	9 14	9 27	9 38	9 48	9 59	10 10	10 23	10 31	10 39
19	8 33	8 54	9 10	9 23	9 35	9 44	10 02	10 17	10 31	10 45	11 01	11 19	11 29	11 41
20	8 46	9 13	9 34	9 51	10 06	10 18	10 40	10 59	11 16	11 34	11 53	12 15	12 29	12 44
21	9 05	9 40	10 05	10 26	10 43	10 58	11 23	11 45	12 05	12 25	12 48	13 13	13 29	13 46
22	9 36	10 16	10 45	11 08	11 27	11 43	12 11	12 35	12 57	13 19	13 43	14 11	14 27	14 47
23	10 22	11 05	11 36	11 59	12 19	12 36	13 04	13 28	13 51	14 14	14 38	15 06	15 23	15 43
24	11 29	12 09	12 38	13 00	13 19	13 35	14 02	14 25	14 47	15 08	15 31	15 58	16 14	16 32
25	12 52	13 25	13 49	14 09	14 25	14 39	15 02	15 23	15 42	16 01	16 21	16 44	16 58	17 14
26	14 23	14 47	15 06	15 21	15 34	15 45	16 04	16 20	16 36	16 51	17 07	17 26	17 36	17 49
27	15 57	16 13	16 25	16 36	16 44	16 52	17 05	17 17	17 28	17 38	17 50	18 03	18 10	18 19
28	17 30	17 38	17 44	17 50	17 55	17 59	18 06	18 12	18 18	18 23	18 30	18 37	18 41	18 45
29	19 03	19 03	19 04	19 04	19 05	19 05	19 06	19 06	19 07	19 08	19 08	19 09	19 09	19 10
30	20 36	20 29	20 24	20 19	20 15	20 12	20 06	20 01	19 56	19 52	19 47	19 41	19 38	19 35
31	22 11	21 57	21 45	21 35	21 27	21 20	21 08	20 57	20 47	20 38	20 27	20 16	20 09	20 02
Sept. 1	23 49	23 26	23 08	22 53	22 41	22 30	22 12	21 56	21 41	21 27	21 11	20 54	20 43	20 32
2	.. ..	.. ..	.. ..	.. ..	23 54	23 40	23 17	22 57	22 38	22 19	21 59	21 37	21 23	21 08
3	1 27	0 54	0 30	0 10	.. ..	.. ..	.. ..	23 59	23 38	23 16	22 53	22 26	22 11	21 53
4	2 57	2 17	1 47	1 25	1 06	0 50	0 22	.. ..	.. ..	.. ..	23 51	23 23	23 06	22 47
5	4 10	3 26	2 55	2 31	2 11	1 54	1 26	1 01	0 38	0 16	.. ..	.. ..	.. ..	23 49
6	4 59	4 19	3 50	3 27	3 08	2 52	2 24	2 01	1 39	1 16	0 53	0 25	0 09	.. ..
7	5 29	4 56	4 32	4 12	3 55	3 41	3 17	2 55	2 36	2 16	1 54	1 29	1 15	0 58
8	5 47	5 23	5 03	4 48	4 34	4 22	4 02	3 45	3 28	3 12	2 54	2 34	2 21	2 08
9	5 59	5 42	5 28	5 16	5 07	4 58	4 43	4 29	4 17	4 04	3 51	3 35	3 26	3 16

MOONSET

Lat.	−55°	−50°	−45°	−40°	−35°	−30°	−20°	−10°	0°	+10°	+20°	+30°	+35°	+40°
	h m	h m	h m	h m	h m	h m	h m	h m	h m	h m	h m	h m	h m	h m
Aug. 16	21 22	21 16	21 11	21 07	21 04	21 01	20 56	20 51	20 47	20 42	20 38	20 33	20 30	20 27
17	22 38	22 25	22 16	22 08	22 01	21 55	21 45	21 35	21 27	21 19	21 10	21 00	20 54	20 47
18	23 54	23 36	23 21	23 09	22 59	22 50	22 34	22 21	22 09	21 56	21 43	21 28	21 20	21 10
19	.. ..	.. ..	.. ..	.. ..	23 58	23 46	23 26	23 09	22 53	22 37	22 19	22 00	21 49	21 36
20	1 13	0 47	0 27	0 11	.. ..	.. ..	.. ..	23 59	23 39	23 20	23 00	22 36	22 22	22 06
21	2 32	1 58	1 33	1 14	0 57	0 43	0 19	.. ..	.. ..	.. ..	23 45	23 18	23 03	22 44
22	3 47	3 07	2 38	2 16	1 57	1 41	1 14	0 51	0 30	0 08	.. ..	.. ..	23 51	23 31
23	4 51	4 08	3 38	3 14	2 55	2 38	2 10	1 45	1 23	1 00	0 36	0 07	.. ..	.. ..
24	5 39	4 59	4 30	4 07	3 48	3 31	3 04	2 40	2 18	1 55	1 31	1 03	0 47	0 27
25	6 12	5 38	5 12	4 52	4 35	4 20	3 55	3 34	3 13	2 53	2 31	2 05	1 50	1 32
26	6 33	6 07	5 47	5 30	5 16	5 04	4 43	4 25	4 08	3 51	3 32	3 10	2 58	2 43
27	6 47	6 29	6 14	6 02	5 52	5 43	5 27	5 14	5 01	4 48	4 33	4 17	4 08	3 57
28	6 57	6 46	6 38	6 30	6 24	6 18	6 09	6 00	5 52	5 43	5 34	5 24	5 18	5 11
29	7 06	7 02	6 59	6 56	6 53	6 51	6 48	6 44	6 41	6 38	6 34	6 30	6 28	6 25
30	7 14	7 16	7 19	7 20	7 22	7 24	7 26	7 28	7 30	7 32	7 34	7 37	7 38	7 40
31	7 22	7 32	7 39	7 46	7 52	7 57	8 05	8 13	8 20	8 28	8 35	8 44	8 49	8 55
Sept. 1	7 32	7 49	8 03	8 14	8 24	8 32	8 47	9 00	9 13	9 25	9 38	9 53	10 02	10 12
2	7 46	8 12	8 31	8 47	9 01	9 13	9 33	9 51	10 08	10 25	10 43	11 03	11 16	11 30
3	8 08	8 42	9 07	9 27	9 44	9 59	10 24	10 46	11 06	11 26	11 48	12 14	12 29	12 46
4	8 42	9 23	9 53	10 16	10 36	10 52	11 20	11 44	12 07	12 29	12 53	13 21	13 38	13 57
5	9 36	10 20	10 51	11 15	11 35	11 52	12 20	12 45	13 07	13 30	13 54	14 23	14 39	14 59
6	10 51	11 31	11 59	12 21	12 40	12 56	13 22	13 45	14 06	14 27	14 50	15 16	15 31	15 49
7	12 17	12 49	13 12	13 31	13 47	14 00	14 23	14 43	15 01	15 20	15 39	16 01	16 14	16 29
8	13 45	14 08	14 26	14 41	14 53	15 04	15 22	15 38	15 52	16 07	16 22	16 40	16 50	17 01
9	15 10	15 26	15 38	15 48	15 57	16 04	16 17	16 28	16 39	16 49	17 00	17 13	17 20	17 28

.. .. indicates phenomenon will occur the next day.

UNIVERSAL TIME FOR MERIDIAN OF GREENWICH

MOONRISE

Lat.	+40°	+42°	+44°	+46°	+48°	+50°	+52°	+54°	+56°	+58°	+60°	+62°	+64°	+66°
	h m	h m	h m	h m	h m	h m	h m	h m	h m	h m	h m	h m	h m	h m
Aug. 16	8 38	8 39	8 40	8 40	8 41	8 42	8 44	8 45	8 46	8 48	8 50	8 52	8 54	8 56
17	9 38	9 41	9 43	9 46	9 49	9 52	9 55	9 59	10 03	10 08	10 13	10 19	10 26	10 34
18	10 39	10 43	10 47	10 52	10 56	11 02	11 07	11 14	11 21	11 29	11 38	11 49	12 01	12 17
19	11 41	11 46	11 52	11 58	12 05	12 13	12 21	12 30	12 41	12 52	13 06	13 23	13 44	14 11
20	12 44	12 50	12 58	13 06	13 15	13 24	13 35	13 47	14 02	14 18	14 39	15 05	15 41	17 04
21	13 46	13 54	14 03	14 13	14 23	14 35	14 48	15 04	15 22	15 44	16 13	16 57	■	■
22	14 47	14 56	15 05	15 16	15 27	15 41	15 56	16 13	16 34	17 01	17 40	■	■	■
23	15 43	15 52	16 01	16 12	16 24	16 37	16 52	17 10	17 32	17 59	18 38	■	■	■
24	16 32	16 40	16 49	16 59	17 10	17 22	17 36	17 51	18 10	18 32	19 02	19 45	■	■
25	17 14	17 21	17 28	17 37	17 46	17 55	18 07	18 19	18 33	18 50	19 10	19 35	20 10	21 10
26	17 49	17 54	18 00	18 06	18 13	18 21	18 29	18 38	18 48	19 00	19 13	19 29	19 48	20 12
27	18 19	18 22	18 26	18 30	18 35	18 40	18 46	18 52	18 58	19 06	19 14	19 23	19 35	19 48
28	18 45	18 47	18 49	18 51	18 54	18 56	18 59	19 02	19 06	19 10	19 14	19 19	19 24	19 30
29	19 10	19 10	19 10	19 11	19 11	19 11	19 12	19 12	19 12	19 13	19 13	19 14	19 14	19 15
30	19 35	19 33	19 32	19 30	19 28	19 26	19 24	19 21	19 19	19 16	19 13	19 09	19 05	19 00
31	20 02	19 58	19 55	19 51	19 47	19 42	19 38	19 32	19 26	19 20	19 13	19 04	18 55	18 44
Sept. 1	20 32	20 27	20 21	20 15	20 09	20 02	19 54	19 46	19 37	19 26	19 14	18 59	18 42	18 22
2	21 08	21 01	20 54	20 46	20 37	20 28	20 17	20 05	19 52	19 36	19 17	18 54	18 24	17 37
3	21 53	21 44	21 36	21 26	21 15	21 03	20 50	20 35	20 17	19 55	19 27	18 47	▭	▭
4	22 47	22 38	22 28	22 18	22 06	21 53	21 38	21 20	20 59	20 32	19 55	▭	▭	▭
5	23 49	23 41	23 31	23 21	23 09	22 57	22 42	22 25	22 04	21 38	21 03	19 46	▭	▭
6	.. ..	.. ..	.. ..	.. ..	.. ..	.. ..	23 59	23 44	23 27	23 07	22 41	22 05	20 45	▭
7	0 58	0 50	0 42	0 33	0 23	0 11	.. ..	.. ..	.. ..	.. ..	.. ..	.. ..	23 35	22 53
8	2 08	2 01	1 55	1 47	1 39	1 30	1 21	1 10	0 57	0 42	0 25	0 03	.. ..	.. ..
9	3 16	3 11	3 06	3 01	2 55	2 49	2 42	2 34	2 25	2 16	2 04	1 51	1 34	1 14

MOONSET

Lat.	+40°	+42°	+44°	+46°	+48°	+50°	+52°	+54°	+56°	+58°	+60°	+62°	+64°	+66°
	h m	h m	h m	h m	h m	h m	h m	h m	h m	h m	h m	h m	h m	h m
Aug. 16	20 27	20 25	20 24	20 22	20 20	20 18	20 16	20 14	20 11	20 08	20 05	20 02	19 58	19 53
17	20 47	20 44	20 41	20 38	20 34	20 30	20 26	20 22	20 16	20 11	20 04	19 57	19 49	19 39
18	21 10	21 05	21 01	20 56	20 50	20 44	20 38	20 31	20 23	20 14	20 04	19 52	19 39	19 22
19	21 36	21 30	21 24	21 17	21 10	21 02	20 53	20 43	20 32	20 19	20 04	19 47	19 25	18 58
20	22 06	21 59	21 52	21 43	21 34	21 24	21 13	21 00	20 45	20 28	20 07	19 41	19 03	17 40
21	22 44	22 36	22 27	22 17	22 07	21 55	21 41	21 25	21 07	20 44	20 15	19 31	■	■
22	23 31	23 22	23 12	23 02	22 50	22 37	22 22	22 04	21 43	21 16	20 37	■	■	■
23	.. ..	.. ..	.. ..	23 58	23 46	23 33	23 18	23 01	22 39	22 12	21 33	■	■	■
24	0 27	0 18	0 09	.. ..	.. ..	.. ..	.. ..	.. ..	23 57	23 35	23 06	22 22	■	■
25	1 32	1 24	1 15	1 06	0 55	0 44	0 30	0 15	.. ..	.. ..	.. ..	.. ..	23 53	22 54
26	2 43	2 36	2 29	2 21	2 13	2 03	1 53	1 41	1 27	1 11	0 52	0 27	.. ..	.. ..
27	3 57	3 52	3 46	3 41	3 35	3 28	3 20	3 12	3 02	2 52	2 39	2 24	2 07	1 44
28	5 11	5 08	5 05	5 01	4 57	4 53	4 49	4 44	4 38	4 32	4 25	4 16	4 07	3 55
29	6 25	6 24	6 23	6 22	6 20	6 19	6 17	6 15	6 13	6 10	6 08	6 05	6 01	5 57
30	7 40	7 41	7 41	7 42	7 43	7 44	7 45	7 46	7 47	7 49	7 50	7 52	7 54	7 56
31	8 55	8 58	9 01	9 04	9 07	9 11	9 14	9 19	9 23	9 29	9 35	9 41	9 49	9 59
Sept. 1	10 12	10 17	10 22	10 27	10 33	10 39	10 46	10 53	11 02	11 11	11 23	11 36	11 52	12 11
2	11 30	11 36	11 43	11 51	11 59	12 08	12 18	12 29	12 42	12 57	13 15	13 37	14 07	14 52
3	12 46	12 54	13 03	13 12	13 22	13 34	13 47	14 02	14 19	14 41	15 08	15 48	▭	▭
4	13 57	14 06	14 15	14 26	14 38	14 51	15 06	15 23	15 44	16 10	16 47	▭	▭	▭
5	14 59	15 07	15 17	15 27	15 39	15 52	16 07	16 24	16 45	17 11	17 47	19 03	▭	▭
6	15 49	15 57	16 06	16 15	16 25	16 37	16 50	17 05	17 22	17 43	18 09	18 46	20 07	▭
7	16 29	16 36	16 43	16 51	16 59	17 09	17 19	17 31	17 44	17 59	18 17	18 40	19 09	19 52
8	17 01	17 06	17 12	17 18	17 24	17 31	17 39	17 47	17 57	18 08	18 20	18 35	18 52	19 13
9	17 28	17 31	17 35	17 39	17 44	17 48	17 54	17 59	18 06	18 13	18 21	18 30	18 40	18 53

▭ indicates Moon continuously above horizon.
■ indicates Moon continuously below horizon.
.. .. indicates phenomenon will occur the next day.

UNIVERSAL TIME FOR MERIDIAN OF GREENWICH

MOONRISE

Lat.	−55°	−50°	−45°	−40°	−35°	−30°	−20°	−10°	0°	+10°	+20°	+30°	+35°	+40°
	h m	h m	h m	h m	h m	h m	h m	h m	h m	h m	h m	h m	h m	h m
Sept. 8	5 47	5 23	5 03	4 48	4 34	4 22	4 02	3 45	3 28	3 12	2 54	2 34	2 21	2 08
9	5 59	5 42	5 28	5 16	5 07	4 58	4 43	4 29	4 17	4 04	3 51	3 35	3 26	3 16
10	6 08	5 57	5 48	5 41	5 34	5 29	5 19	5 10	5 02	4 53	4 45	4 35	4 29	4 22
11	6 15	6 10	6 06	6 02	5 59	5 57	5 52	5 48	5 44	5 40	5 36	5 31	5 28	5 25
12	6 21	6 21	6 22	6 22	6 23	6 23	6 23	6 24	6 24	6 25	6 25	6 26	6 26	6 27
13	6 27	6 33	6 38	6 42	6 46	6 49	6 55	7 00	7 04	7 09	7 14	7 20	7 24	7 27
14	6 34	6 45	6 55	7 03	7 10	7 16	7 27	7 36	7 45	7 54	8 03	8 14	8 21	8 28
15	6 42	7 00	7 14	7 26	7 36	7 45	8 00	8 14	8 27	8 40	8 54	9 10	9 19	9 30
16	6 53	7 18	7 37	7 52	8 06	8 17	8 37	8 55	9 11	9 27	9 45	10 06	10 18	10 32
17	7 09	7 41	8 05	8 24	8 40	8 54	9 18	9 38	9 58	10 17	10 38	11 03	11 17	11 34
18	7 34	8 12	8 40	9 02	9 21	9 36	10 03	10 26	10 48	11 09	11 33	12 00	12 16	12 35
19	8 13	8 55	9 25	9 49	10 08	10 25	10 53	11 18	11 40	12 03	12 27	12 55	13 12	13 32
20	9 10	9 52	10 21	10 45	11 04	11 20	11 48	12 12	12 34	12 56	13 20	13 48	14 04	14 23
21	10 25	11 01	11 27	11 48	12 06	12 21	12 46	13 08	13 28	13 49	14 10	14 35	14 50	15 07
22	11 51	12 19	12 41	12 58	13 12	13 25	13 46	14 05	14 22	14 39	14 57	15 18	15 30	15 44
23	13 22	13 42	13 58	14 10	14 21	14 31	14 47	15 01	15 14	15 27	15 41	15 56	16 05	16 16
24	14 55	15 07	15 16	15 24	15 31	15 37	15 47	15 56	16 04	16 13	16 21	16 31	16 37	16 44
25	16 28	16 32	16 36	16 39	16 41	16 44	16 47	16 51	16 54	16 57	17 01	17 05	17 07	17 09
26	18 02	17 59	17 57	17 55	17 53	17 51	17 49	17 46	17 44	17 42	17 40	17 38	17 36	17 35
27	19 39	19 28	19 19	19 12	19 06	19 01	18 51	18 43	18 36	18 28	18 21	18 12	18 07	18 01
28	21 19	21 00	20 44	20 32	20 21	20 12	19 56	19 43	19 30	19 17	19 04	18 49	18 40	18 31
29	23 01	22 32	22 10	21 52	21 38	21 25	21 03	20 45	20 28	20 10	19 52	19 31	19 19	19 06
30	.. ..	.. ..	23 33	23 11	22 53	22 37	22 11	21 49	21 28	21 08	20 46	20 20	20 06	19 48
Oct. 1	0 38	0 00	.. ..	.. ..	.. ..	23 46	23 18	22 53	22 31	22 08	21 44	21 16	21 00	20 41
2	2 00	1 17	0 46	0 22	0 03	.. ..	.. ..	23 55	23 33	23 10	22 46	22 18	22 01	21 42

MOONSET

Lat.	−55°	−50°	−45°	−40°	−35°	−30°	−20°	−10°	0°	+10°	+20°	+30°	+35°	+40°
	h m	h m	h m	h m	h m	h m	h m	h m	h m	h m	h m	h m	h m	h m
Sept. 8	13 45	14 08	14 26	14 41	14 53	15 04	15 22	15 38	15 52	16 07	16 22	16 40	16 50	17 01
9	15 10	15 26	15 38	15 48	15 57	16 04	16 17	16 28	16 39	16 49	17 00	17 13	17 20	17 28
10	16 31	16 40	16 47	16 53	16 58	17 02	17 10	17 16	17 22	17 28	17 35	17 42	17 46	17 51
11	17 49	17 52	17 54	17 55	17 57	17 58	18 00	18 02	18 03	18 05	18 07	18 09	18 10	18 11
12	19 05	19 01	18 59	18 56	18 54	18 52	18 49	18 46	18 44	18 41	18 38	18 35	18 33	18 31
13	20 21	20 11	20 03	19 57	19 51	19 46	19 38	19 31	19 24	19 17	19 10	19 02	18 57	18 51
14	21 37	21 21	21 08	20 58	20 49	20 41	20 28	20 16	20 05	19 54	19 42	19 29	19 22	19 13
15	22 55	22 32	22 14	21 59	21 47	21 37	21 18	21 03	20 48	20 33	20 18	20 00	19 49	19 38
16	.. ..	23 43	23 20	23 02	22 46	22 33	22 11	21 51	21 33	21 15	20 56	20 34	20 21	20 06
17	0 13	.. ..	.. ..	.. ..	23 46	23 30	23 05	22 42	22 22	22 01	21 39	21 13	20 58	20 41
18	1 30	0 52	0 25	0 04	.. ..	.. ..	23 59	23 35	23 13	22 50	22 27	21 59	21 42	21 23
19	2 38	1 56	1 26	1 03	0 43	0 27	.. ..	.. ..	.. ..	23 43	23 19	22 51	22 34	22 14
20	3 32	2 50	2 20	1 57	1 38	1 21	0 53	0 29	0 06	.. ..	.. ..	23 49	23 33	23 14
21	4 10	3 33	3 06	2 44	2 26	2 11	1 44	1 22	1 00	0 39	0 16	.. ..	.. ..	.. ..
22	4 35	4 06	3 43	3 25	3 09	2 56	2 33	2 13	1 54	1 35	1 15	0 51	0 37	0 21
23	4 52	4 30	4 13	3 59	3 47	3 36	3 18	3 02	2 47	2 32	2 15	1 56	1 45	1 32
24	5 04	4 50	4 38	4 28	4 20	4 13	4 00	3 49	3 38	3 27	3 16	3 02	2 54	2 45
25	5 13	5 06	5 00	4 55	4 51	4 47	4 40	4 34	4 28	4 22	4 16	4 09	4 04	4 00
26	5 22	5 21	5 21	5 20	5 20	5 19	5 19	5 18	5 18	5 17	5 16	5 16	5 15	5 15
27	5 30	5 36	5 41	5 46	5 49	5 53	5 58	6 04	6 08	6 13	6 18	6 24	6 27	6 31
28	5 40	5 53	6 04	6 13	6 21	6 28	6 40	6 51	7 01	7 11	7 22	7 34	7 41	7 50
29	5 52	6 14	6 31	6 45	6 57	7 08	7 26	7 42	7 57	8 12	8 28	8 47	8 57	9 10
30	6 11	6 42	7 05	7 24	7 40	7 53	8 17	8 37	8 56	9 15	9 36	10 00	10 14	10 30
Oct. 1	6 41	7 20	7 49	8 11	8 30	8 46	9 13	9 36	9 58	10 20	10 43	11 11	11 27	11 46
2	7 30	8 13	8 44	9 08	9 28	9 45	10 13	10 38	11 01	11 23	11 48	12 16	12 33	12 52

.. .. indicates phenomenon will occur the next day.

UNIVERSAL TIME FOR MERIDIAN OF GREENWICH

MOONRISE

Lat.	+40°	+42°	+44°	+46°	+48°	+50°	+52°	+54°	+56°	+58°	+60°	+62°	+64°	+66°
	h m	h m	h m	h m	h m	h m	h m	h m	h m	h m	h m	h m	h m	h m
Sept. 8	2 08	2 01	1 55	1 47	1 39	1 30	1 21	1 10	0 57	0 42	0 25	0 03	.. ..	.. ..
9	3 16	3 11	3 06	3 01	2 55	2 49	2 42	2 34	2 25	2 16	2 04	1 51	1 34	1 14
10	4 22	4 19	4 16	4 12	4 09	4 05	4 00	3 55	3 50	3 44	3 37	3 29	3 20	3 09
11	5 25	5 24	5 22	5 21	5 19	5 17	5 15	5 13	5 11	5 08	5 05	5 01	4 58	4 53
12	6 27	6 27	6 27	6 27	6 28	6 28	6 28	6 28	6 29	6 29	6 30	6 30	6 31	6 31
13	7 27	7 29	7 31	7 33	7 35	7 37	7 40	7 43	7 46	7 49	7 53	7 58	8 03	8 09
14	8 28	8 31	8 35	8 39	8 43	8 47	8 52	8 57	9 03	9 10	9 18	9 26	9 37	9 49
15	9 30	9 34	9 39	9 45	9 51	9 58	10 05	10 13	10 22	10 32	10 44	10 59	11 16	11 37
16	10 32	10 38	10 45	10 52	11 00	11 09	11 19	11 30	11 42	11 57	12 15	12 37	13 05	13 48
17	11 34	11 42	11 50	11 59	12 09	12 20	12 32	12 46	13 03	13 23	13 48	14 23	15 30	■
18	12 35	12 43	12 53	13 03	13 14	13 27	13 41	13 58	14 18	14 43	15 18	16 26	■	■
19	13 32	13 41	13 51	14 01	14 13	14 27	14 42	15 00	15 21	15 49	16 29	■	■	■
20	14 23	14 32	14 41	14 51	15 03	15 15	15 30	15 47	16 07	16 31	17 05	18 05	■	■
21	15 07	15 15	15 23	15 32	15 42	15 53	16 05	16 19	16 35	16 55	17 19	17 50	18 42	■
22	15 44	15 50	15 57	16 04	16 12	16 21	16 30	16 41	16 53	17 07	17 24	17 43	18 08	18 42
23	16 16	16 20	16 25	16 31	16 36	16 43	16 49	16 57	17 05	17 15	17 25	17 38	17 52	18 10
24	16 44	16 47	16 50	16 53	16 56	17 00	17 04	17 09	17 14	17 19	17 26	17 33	17 41	17 50
25	17 09	17 10	17 12	17 13	17 14	17 16	17 17	17 19	17 21	17 23	17 25	17 28	17 31	17 34
26	17 35	17 34	17 33	17 32	17 32	17 31	17 30	17 29	17 28	17 26	17 25	17 23	17 22	17 20
27	18 01	17 59	17 56	17 53	17 50	17 47	17 43	17 39	17 35	17 30	17 25	17 19	17 12	17 04
28	18 31	18 26	18 22	18 17	18 11	18 06	17 59	17 52	17 44	17 36	17 26	17 14	17 01	16 45
29	19 06	18 59	18 53	18 46	18 38	18 30	18 20	18 10	17 58	17 44	17 28	17 10	16 46	16 15
30	19 48	19 41	19 32	19 23	19 13	19 02	18 50	18 36	18 19	18 00	17 36	17 04	16 13	▭
Oct. 1	20 41	20 32	20 22	20 12	20 01	19 48	19 33	19 16	18 56	18 31	17 57	16 55	▭	▭
2	21 42	21 33	21 24	21 13	21 01	20 48	20 33	20 16	19 55	19 29	18 51	▭	▭	▭

MOONSET

Lat.	+40°	+42°	+44°	+46°	+48°	+50°	+52°	+54°	+56°	+58°	+60°	+62°	+64°	+66°
	h m	h m	h m	h m	h m	h m	h m	h m	h m	h m	h m	h m	h m	h m
Sept. 8	17 01	17 06	17 12	17 18	17 24	17 31	17 39	17 47	17 57	18 08	18 20	18 35	18 52	19 13
9	17 28	17 31	17 35	17 39	17 44	17 48	17 54	17 59	18 06	18 13	18 21	18 30	18 40	18 53
10	17 51	17 53	17 55	17 57	18 00	18 02	18 05	18 08	18 12	18 16	18 20	18 25	18 31	18 37
11	18 11	18 12	18 12	18 13	18 14	18 15	18 15	18 16	18 17	18 18	18 19	18 21	18 22	18 24
12	18 31	18 30	18 29	18 28	18 27	18 26	18 25	18 23	18 22	18 20	18 18	18 16	18 14	18 11
13	18 51	18 49	18 47	18 44	18 41	18 38	18 35	18 31	18 27	18 22	18 17	18 11	18 05	17 57
14	19 13	19 09	19 05	19 01	18 56	18 51	18 46	18 39	18 33	18 25	18 16	18 07	17 55	17 42
15	19 38	19 32	19 27	19 21	19 14	19 07	18 59	18 50	18 41	18 29	18 17	18 02	17 44	17 21
16	20 06	20 00	19 52	19 45	19 36	19 27	19 17	19 05	18 52	18 37	18 18	17 56	17 27	16 43
17	20 41	20 33	20 25	20 15	20 05	19 54	19 41	19 27	19 10	18 49	18 24	17 48	16 40	■
18	21 23	21 14	21 05	20 55	20 43	20 30	20 16	19 59	19 38	19 13	18 38	17 30	■	■
19	22 14	22 05	21 56	21 45	21 33	21 20	21 04	20 47	20 25	19 57	19 17	■	■	■
20	23 14	23 06	22 57	22 46	22 35	22 23	22 09	21 52	21 32	21 08	20 34	19 34	■	■
21	.. ..	.. ..	.. ..	23 57	23 48	23 37	23 25	23 12	22 56	22 37	22 13	21 42	20 51	■
22	0 21	0 14	0 06	.. ..	.. ..	.. ..	.. ..	.. ..	.. ..	.. ..	23 59	23 40	23 16	22 43
23	1 32	1 26	1 20	1 13	1 06	0 58	0 49	0 39	0 28	0 14	.. ..	.. ..	.. ..	.. ..
24	2 45	2 41	2 37	2 32	2 27	2 22	2 16	2 09	2 02	1 53	1 44	1 33	1 19	1 03
25	4 00	3 57	3 55	3 53	3 50	3 47	3 44	3 41	3 37	3 32	3 28	3 22	3 16	3 08
26	5 15	5 14	5 14	5 14	5 14	5 13	5 13	5 13	5 12	5 12	5 11	5 10	5 10	5 09
27	6 31	6 33	6 35	6 37	6 39	6 41	6 44	6 46	6 49	6 53	6 56	7 01	7 06	7 12
28	7 50	7 53	7 57	8 02	8 06	8 11	8 17	8 23	8 29	8 37	8 46	8 56	9 08	9 22
29	9 10	9 16	9 22	9 28	9 35	9 43	9 52	10 02	10 13	10 25	10 40	10 58	11 21	11 51
30	10 30	10 37	10 45	10 54	11 04	11 14	11 26	11 40	11 56	12 14	12 38	13 09	14 00	▭
Oct. 1	11 46	11 54	12 04	12 14	12 25	12 38	12 52	13 09	13 29	13 54	14 27	15 29	▭	▭
2	12 52	13 01	13 11	13 21	13 33	13 46	14 01	14 19	14 40	15 06	15 43	▭	▭	▭

▭ indicates Moon continuously above horizon.
■ indicates Moon continuously below horizon.
.. .. indicates phenomenon will occur the next day.

UNIVERSAL TIME FOR MERIDIAN OF GREENWICH

MOONRISE

Lat.	−55°	−50°	−45°	−40°	−35°	−30°	−20°	−10°	0°	+10°	+20°	+30°	+35°	+40°
	h m	h m	h m	h m	h m	h m	h m	h m	h m	h m	h m	h m	h m	h m
Oct. 1	0 38	0 00	.. ..	.. ..	.. ..	23 46	23 18	22 53	22 31	22 08	21 44	21 16	21 00	20 41
2	2 00	1 17	0 46	0 22	0 03	.. ..	.. ..	23 55	23 33	23 10	22 46	22 18	22 01	21 42
3	2 58	2 16	1 46	1 23	1 04	0 47	0 19	.. ..	.. ..	.. ..	23 48	23 23	23 07	22 50
4	3 33	2 58	2 32	2 12	1 54	1 39	1 14	0 52	0 31	0 11	.. ..	.. ..	.. ..	23 59
5	3 55	3 28	3 07	2 50	2 36	2 23	2 02	1 43	1 25	1 08	0 49	0 27	0 14	.. ..
6	4 08	3 49	3 33	3 20	3 09	3 00	2 43	2 29	2 15	2 01	1 46	1 29	1 19	1 08
7	4 18	4 05	3 54	3 46	3 38	3 32	3 20	3 10	3 00	2 51	2 40	2 29	2 22	2 14
8	4 25	4 18	4 13	4 08	4 04	4 00	3 54	3 48	3 43	3 37	3 32	3 25	3 22	3 17
9	4 31	4 30	4 29	4 28	4 27	4 27	4 25	4 24	4 23	4 22	4 21	4 20	4 19	4 19
10	4 37	4 42	4 45	4 48	4 50	4 53	4 56	5 00	5 03	5 06	5 10	5 14	5 16	5 19
11	4 44	4 54	5 02	5 08	5 14	5 19	5 28	5 36	5 43	5 51	5 59	6 08	6 13	6 19
12	4 52	5 07	5 20	5 30	5 39	5 47	6 01	6 13	6 24	6 36	6 48	7 03	7 11	7 20
13	5 02	5 24	5 41	5 56	6 08	6 18	6 37	6 53	7 08	7 23	7 39	7 58	8 09	8 22
14	5 16	5 45	6 07	6 25	6 40	6 53	7 16	7 35	7 54	8 12	8 32	8 55	9 09	9 24
15	5 38	6 14	6 40	7 01	7 18	7 34	7 59	8 22	8 42	9 03	9 26	9 52	10 08	10 26
16	6 11	6 52	7 21	7 44	8 03	8 20	8 47	9 11	9 33	9 56	10 20	10 48	11 04	11 24
17	7 00	7 43	8 12	8 36	8 55	9 12	9 40	10 04	10 26	10 48	11 12	11 40	11 57	12 16
18	8 07	8 46	9 13	9 35	9 53	10 09	10 35	10 58	11 19	11 40	12 02	12 29	12 44	13 02
19	9 27	9 58	10 22	10 41	10 56	11 10	11 33	11 53	12 11	12 30	12 49	13 12	13 25	13 40
20	10 53	11 17	11 35	11 49	12 02	12 13	12 31	12 47	13 02	13 17	13 33	13 51	14 01	14 13
21	12 22	12 38	12 50	13 00	13 09	13 17	13 30	13 41	13 52	14 02	14 13	14 26	14 33	14 42
22	13 52	14 00	14 07	14 12	14 17	14 21	14 28	14 34	14 40	14 46	14 52	14 59	15 03	15 08
23	15 24	15 25	15 25	15 26	15 27	15 27	15 28	15 28	15 29	15 30	15 31	15 31	15 32	15 33
24	16 59	16 52	16 46	16 42	16 38	16 35	16 29	16 24	16 19	16 15	16 10	16 05	16 02	15 58
25	18 37	18 22	18 11	18 01	17 53	17 46	17 33	17 23	17 13	17 03	16 52	16 41	16 34	16 26

MOONSET

Lat.	−55°	−50°	−45°	−40°	−35°	−30°	−20°	−10°	0°	+10°	+20°	+30°	+35°	+40°
	h m	h m	h m	h m	h m	h m	h m	h m	h m	h m	h m	h m	h m	h m
Oct. 1	6 41	7 20	7 49	8 11	8 30	8 46	9 13	9 36	9 58	10 20	10 43	11 11	11 27	11 46
2	7 30	8 13	8 44	9 08	9 28	9 45	10 13	10 38	11 01	11 23	11 48	12 16	12 33	12 52
3	8 40	9 21	9 50	10 13	10 32	10 49	11 16	11 39	12 01	12 23	12 46	13 13	13 29	13 47
4	10 04	10 38	11 03	11 23	11 39	11 54	12 18	12 39	12 58	13 17	13 37	14 01	14 15	14 30
5	11 32	11 57	12 17	12 33	12 46	12 57	13 17	13 34	13 50	14 05	14 22	14 41	14 52	15 05
6	12 57	13 15	13 29	13 40	13 50	13 58	14 13	14 25	14 37	14 49	15 01	15 15	15 23	15 32
7	14 18	14 29	14 38	14 45	14 51	14 56	15 05	15 13	15 21	15 28	15 36	15 45	15 50	15 56
8	15 36	15 40	15 44	15 47	15 50	15 52	15 56	15 59	16 02	16 05	16 09	16 12	16 15	16 17
9	16 52	16 50	16 49	16 48	16 47	16 46	16 45	16 44	16 42	16 41	16 40	16 39	16 38	16 37
10	18 07	17 59	17 53	17 48	17 44	17 40	17 33	17 28	17 22	17 17	17 11	17 05	17 01	16 57
11	19 22	19 09	18 58	18 49	18 41	18 34	18 23	18 12	18 03	17 53	17 43	17 32	17 25	17 18
12	20 40	20 19	20 03	19 50	19 39	19 29	19 13	18 58	18 45	18 32	18 18	18 01	17 52	17 41
13	21 58	21 30	21 09	20 52	20 38	20 26	20 05	19 47	19 30	19 13	18 55	18 34	18 22	18 09
14	23 15	22 40	22 14	21 54	21 37	21 23	20 58	20 37	20 17	19 57	19 36	19 12	18 58	18 41
15	.. ..	23 46	23 17	22 54	22 35	22 19	21 52	21 29	21 07	20 45	20 22	19 55	19 39	19 20
16	0 26	.. ..	.. ..	23 49	23 30	23 13	22 45	22 21	21 59	21 36	21 12	20 44	20 27	20 08
17	1 25	0 43	0 13	.. ..	.. ..	.. ..	23 37	23 13	22 52	22 30	22 06	21 39	21 22	21 03
18	2 08	1 29	1 01	0 38	0 20	0 04	.. ..	.. ..	23 44	23 24	23 03	22 38	22 23	22 06
19	2 37	2 04	1 40	1 20	1 04	0 50	0 25	0 04	.. ..	.. ..	.. ..	23 40	23 27	23 13
20	2 56	2 31	2 12	1 56	1 42	1 31	1 10	0 52	0 36	0 19	0 01	.. ..	.. ..	.. ..
21	3 10	2 52	2 38	2 26	2 16	2 07	1 52	1 39	1 26	1 13	0 59	0 43	0 34	0 23
22	3 20	3 09	3 01	2 53	2 47	2 41	2 32	2 23	2 15	2 06	1 58	1 47	1 41	1 35
23	3 29	3 25	3 21	3 18	3 16	3 14	3 10	3 07	3 03	3 00	2 56	2 52	2 50	2 47
24	3 37	3 39	3 42	3 43	3 45	3 46	3 49	3 51	3 53	3 55	3 57	3 59	4 00	4 02
25	3 46	3 56	4 03	4 10	4 15	4 21	4 29	4 37	4 44	4 51	4 59	5 08	5 13	5 19

.. .. indicates phenomenon will occur the next day.

UNIVERSAL TIME FOR MERIDIAN OF GREENWICH

MOONRISE

Lat.	+40°	+42°	+44°	+46°	+48°	+50°	+52°	+54°	+56°	+58°	+60°	+62°	+64°	+66°
	h m	h m	h m	h m	h m	h m	h m	h m	h m	h m	h m	h m	h m	h m
Oct. 1	20 41	20 32	20 22	20 12	20 01	19 48	19 33	19 16	18 56	18 31	17 57	16 55	□	□
2	21 42	21 33	21 24	21 13	21 01	20 48	20 33	20 16	19 55	19 29	18 51	□	□	□
3	22 50	22 42	22 33	22 23	22 13	22 01	21 48	21 33	21 14	20 52	20 24	19 41	□	□
4	23 59	23 53	23 46	23 38	23 29	23 20	23 09	22 57	22 43	22 27	22 07	21 43	21 08	20 07
5	.. ..	.. ..	.. ..	.. ..	.. ..	.. ..	.. ..	.. ..	.. ..	.. ..	23 47	23 32	23 13	22 49
6	1 08	1 03	0 57	0 52	0 45	0 38	0 30	0 21	0 12	0 00	.. ..	.. ..	.. ..	.. ..
7	2 14	2 11	2 07	2 03	1 58	1 54	1 49	1 43	1 36	1 29	1 21	1 12	1 01	0 47
8	3 17	3 15	3 13	3 11	3 09	3 06	3 04	3 01	2 57	2 53	2 49	2 44	2 39	2 32
9	4 19	4 18	4 18	4 18	4 17	4 17	4 16	4 16	4 15	4 14	4 14	4 13	4 12	4 11
10	5 19	5 20	5 21	5 23	5 24	5 26	5 28	5 30	5 32	5 34	5 37	5 40	5 43	5 47
11	6 19	6 22	6 25	6 28	6 31	6 35	6 39	6 43	6 48	6 54	7 00	7 07	7 16	7 25
12	7 20	7 25	7 29	7 34	7 39	7 45	7 51	7 58	8 06	8 15	8 26	8 38	8 52	9 09
13	8 22	8 28	8 34	8 41	8 48	8 56	9 05	9 15	9 26	9 39	9 54	10 13	10 36	11 08
14	9 24	9 31	9 39	9 48	9 57	10 07	10 18	10 31	10 46	11 04	11 26	11 55	12 38	■
15	10 26	10 34	10 43	10 52	11 03	11 15	11 29	11 45	12 03	12 27	12 57	13 46	■	■
16	11 24	11 32	11 42	11 52	12 04	12 17	12 32	12 50	13 11	13 38	14 16	■	■	■
17	12 16	12 25	12 34	12 45	12 56	13 09	13 24	13 41	14 02	14 28	15 04	16 23	■	■
18	13 02	13 10	13 18	13 28	13 38	13 50	14 03	14 18	14 36	14 57	15 24	16 02	■	■
19	13 40	13 47	13 54	14 02	14 11	14 21	14 31	14 43	14 57	15 13	15 32	15 56	16 27	17 16
20	14 13	14 18	14 24	14 30	14 37	14 44	14 52	15 01	15 11	15 22	15 35	15 50	16 09	16 31
21	14 42	14 45	14 49	14 54	14 58	15 03	15 08	15 14	15 21	15 28	15 36	15 46	15 57	16 09
22	15 08	15 10	15 12	15 14	15 16	15 19	15 22	15 25	15 28	15 32	15 36	15 41	15 47	15 53
23	15 33	15 33	15 33	15 33	15 34	15 34	15 34	15 35	15 35	15 36	15 36	15 37	15 38	15 39
24	15 58	15 57	15 55	15 53	15 51	15 49	15 47	15 45	15 42	15 40	15 36	15 33	15 29	15 24
25	16 26	16 23	16 19	16 16	16 11	16 07	16 02	15 57	15 51	15 44	15 37	15 29	15 19	15 08

MOONSET

Lat.	+40°	+42°	+44°	+46°	+48°	+50°	+52°	+54°	+56°	+58°	+60°	+62°	+64°	+66°
	h m	h m	h m	h m	h m	h m	h m	h m	h m	h m	h m	h m	h m	h m
Oct. 1	11 46	11 54	12 04	12 14	12 25	12 38	12 52	13 09	13 29	13 54	14 27	15 29	□	□
2	12 52	13 01	13 11	13 21	13 33	13 46	14 01	14 19	14 40	15 06	15 43	□	□	□
3	13 47	13 55	14 04	14 14	14 25	14 37	14 50	15 06	15 24	15 47	16 16	16 59	□	□
4	14 30	14 37	14 45	14 53	15 02	15 12	15 23	15 36	15 50	16 07	16 27	16 53	17 28	18 30
5	15 05	15 10	15 16	15 23	15 30	15 37	15 46	15 55	16 06	16 18	16 31	16 48	17 08	17 33
6	15 32	15 36	15 41	15 45	15 50	15 56	16 02	16 08	16 15	16 24	16 33	16 43	16 56	17 10
7	15 56	15 58	16 01	16 04	16 07	16 10	16 14	16 18	16 22	16 27	16 33	16 39	16 46	16 54
8	16 17	16 18	16 19	16 20	16 22	16 23	16 24	16 26	16 28	16 30	16 32	16 34	16 37	16 41
9	16 37	16 37	16 36	16 36	16 35	16 35	16 34	16 33	16 33	16 32	16 31	16 30	16 29	16 28
10	16 57	16 55	16 53	16 51	16 49	16 46	16 44	16 41	16 38	16 34	16 30	16 26	16 21	16 15
11	17 18	17 15	17 11	17 07	17 03	16 59	16 54	16 49	16 43	16 37	16 30	16 21	16 12	16 01
12	17 41	17 37	17 32	17 26	17 20	17 14	17 07	16 59	16 51	16 41	16 30	16 17	16 01	15 43
13	18 09	18 03	17 56	17 49	17 41	17 33	17 23	17 13	17 01	16 47	16 31	16 12	15 48	15 15
14	18 41	18 34	18 26	18 17	18 07	17 57	17 45	17 32	17 16	16 58	16 35	16 06	15 22	■
15	19 20	19 12	19 03	18 53	18 42	18 30	18 16	18 00	17 41	17 17	16 46	15 58	■	■
16	20 08	19 59	19 49	19 39	19 27	19 14	18 59	18 41	18 20	17 53	17 15	■	■	■
17	21 03	20 55	20 45	20 35	20 23	20 11	19 56	19 39	19 19	18 53	18 17	16 58	■	■
18	22 06	21 58	21 50	21 40	21 30	21 19	21 06	20 51	20 34	20 13	19 47	19 09	■	■
19	23 13	23 07	23 00	22 52	22 44	22 35	22 25	22 13	22 00	21 45	21 27	21 04	20 33	19 45
20	.. ..	.. ..	.. ..	.. ..	.. ..	23 55	23 48	23 40	23 30	23 20	23 08	22 54	22 36	22 15
21	0 23	0 18	0 13	0 08	0 02	.. ..	.. ..	.. ..	.. ..	.. ..	.. ..	.. ..	.. ..	.. ..
22	1 35	1 32	1 28	1 25	1 21	1 17	1 12	1 07	1 02	0 56	0 49	0 41	0 31	0 20
23	2 47	2 46	2 45	2 43	2 42	2 40	2 38	2 37	2 34	2 32	2 29	2 26	2 23	2 18
24	4 02	4 02	4 03	4 04	4 05	4 06	4 07	4 08	4 09	4 10	4 12	4 13	4 15	4 17
25	5 19	5 22	5 25	5 28	5 31	5 34	5 38	5 42	5 47	5 52	5 58	6 05	6 13	6 23

□ indicates Moon continuously above horizon.
■ indicates Moon continuously below horizon.
.. .. indicates phenomenon will occur the next day.

UNIVERSAL TIME FOR MERIDIAN OF GREENWICH

MOONRISE

Lat.	−55°	−50°	−45°	−40°	−35°	−30°	−20°	−10°	0°	+10°	+20°	+30°	+35°	+40°
	h m	h m	h m	h m	h m	h m	h m	h m	h m	h m	h m	h m	h m	h m
Oct. 24	16 59	16 52	16 46	16 42	16 38	16 35	16 29	16 24	16 19	16 15	16 10	16 05	16 02	15 58
25	18 37	18 22	18 11	18 01	17 53	17 46	17 33	17 23	17 13	17 03	16 52	16 41	16 34	16 26
26	20 21	19 56	19 38	19 23	19 10	18 59	18 41	18 25	18 10	17 55	17 39	17 21	17 11	16 59
27	22 04	21 30	21 05	20 45	20 29	20 15	19 51	19 30	19 11	18 52	18 31	18 08	17 55	17 39
28	23 37	22 56	22 26	22 03	21 44	21 28	21 00	20 37	20 15	19 53	19 30	19 03	18 47	18 29
29	.. ..	.. ..	23 35	23 11	22 52	22 35	22 07	21 43	21 20	20 57	20 33	20 05	19 48	19 29
30	0 48	0 05	.. ..	.. ..	23 48	23 33	23 06	22 43	22 22	22 01	21 38	21 11	20 55	20 37
31	1 33	0 55	0 28	0 06	.. ..	.. ..	23 58	23 38	23 20	23 01	22 41	22 18	22 04	21 49
Nov. 1	1 59	1 30	1 07	0 49	0 34	0 21	.. ..	.. ..	.. ..	23 57	23 41	23 22	23 12	22 59
2	2 15	1 54	1 37	1 23	1 11	1 00	0 42	0 27	0 12	.. ..	.. ..	.. ..	.. ..	.. ..
3	2 26	2 12	2 00	1 50	1 41	1 34	1 21	1 10	0 59	0 48	0 36	0 23	0 15	0 07
4	2 34	2 26	2 19	2 13	2 08	2 03	1 56	1 49	1 42	1 36	1 29	1 21	1 16	1 11
5	2 41	2 38	2 36	2 34	2 32	2 30	2 28	2 25	2 23	2 21	2 18	2 16	2 14	2 12
6	2 47	2 50	2 52	2 54	2 55	2 56	2 59	3 01	3 03	3 05	3 07	3 10	3 11	3 13
7	2 54	3 02	3 08	3 14	3 18	3 23	3 30	3 36	3 43	3 49	3 55	4 03	4 07	4 12
8	3 01	3 15	3 26	3 35	3 43	3 50	4 02	4 13	4 23	4 34	4 44	4 57	5 04	5 13
9	3 11	3 31	3 47	4 00	4 11	4 20	4 37	4 52	5 06	5 20	5 35	5 52	6 03	6 14
10	3 24	3 51	4 11	4 28	4 42	4 54	5 15	5 34	5 51	6 08	6 27	6 49	7 02	7 16
11	3 44	4 17	4 42	5 02	5 18	5 33	5 58	6 19	6 39	6 59	7 21	7 46	8 01	8 18
12	4 13	4 52	5 20	5 43	6 01	6 17	6 44	7 08	7 29	7 51	8 15	8 42	8 58	9 17
13	4 57	5 39	6 09	6 32	6 51	7 08	7 35	7 59	8 22	8 44	9 08	9 36	9 52	10 11
14	5 59	6 38	7 06	7 29	7 47	8 03	8 30	8 53	9 14	9 35	9 58	10 25	10 41	10 59
15	7 13	7 47	8 12	8 31	8 48	9 02	9 26	9 47	10 06	10 25	10 46	11 09	11 23	11 39
16	8 36	9 02	9 21	9 37	9 51	10 03	10 23	10 40	10 56	11 12	11 29	11 49	12 00	12 13
17	10 01	10 19	10 34	10 45	10 55	11 04	11 19	11 32	11 44	11 56	12 09	12 24	12 32	12 42

MOONSET

Lat.	−55°	−50°	−45°	−40°	−35°	−30°	−20°	−10°	0°	+10°	+20°	+30°	+35°	+40°
	h m	h m	h m	h m	h m	h m	h m	h m	h m	h m	h m	h m	h m	h m
Oct. 24	3 37	3 39	3 42	3 43	3 45	3 46	3 49	3 51	3 53	3 55	3 57	3 59	4 00	4 02
25	3 46	3 56	4 03	4 10	4 15	4 21	4 29	4 37	4 44	4 51	4 59	5 08	5 13	5 19
26	3 57	4 15	4 28	4 40	4 50	4 58	5 13	5 27	5 39	5 51	6 05	6 20	6 29	6 40
27	4 13	4 39	4 59	5 16	5 30	5 42	6 03	6 21	6 38	6 55	7 14	7 35	7 48	8 02
28	4 39	5 14	5 40	6 00	6 18	6 33	6 58	7 20	7 41	8 02	8 24	8 50	9 05	9 23
29	5 20	6 02	6 32	6 55	7 15	7 31	8 00	8 24	8 46	9 09	9 33	10 01	10 17	10 37
30	6 24	7 07	7 37	8 00	8 20	8 36	9 04	9 28	9 50	10 12	10 36	11 03	11 20	11 38
31	7 47	8 23	8 50	9 11	9 28	9 43	10 08	10 30	10 50	11 10	11 32	11 56	12 11	12 27
Nov. 1	9 16	9 44	10 05	10 22	10 37	10 49	11 10	11 28	11 45	12 02	12 20	12 40	12 52	13 05
2	10 44	11 04	11 19	11 32	11 43	11 52	12 08	12 22	12 35	12 48	13 01	13 17	13 26	13 36
3	12 07	12 19	12 29	12 38	12 45	12 51	13 02	13 11	13 20	13 29	13 38	13 48	13 54	14 01
4	13 25	13 32	13 37	13 41	13 44	13 47	13 53	13 58	14 02	14 06	14 11	14 16	14 19	14 23
5	14 41	14 41	14 41	14 42	14 42	14 42	14 42	14 42	14 42	14 42	14 43	14 43	14 43	14 43
6	15 56	15 50	15 45	15 41	15 38	15 35	15 30	15 26	15 22	15 18	15 14	15 09	15 06	15 03
7	17 10	16 59	16 49	16 42	16 35	16 29	16 19	16 10	16 02	15 54	15 45	15 35	15 30	15 23
8	18 27	18 08	17 54	17 42	17 32	17 24	17 09	16 56	16 44	16 32	16 19	16 04	15 56	15 46
9	19 44	19 19	19 00	18 44	18 31	18 20	18 00	17 43	17 27	17 12	16 55	16 36	16 25	16 12
10	21 02	20 30	20 05	19 46	19 30	19 16	18 53	18 33	18 14	17 55	17 35	17 12	16 58	16 43
11	22 15	21 37	21 09	20 47	20 29	20 13	19 47	19 24	19 03	18 42	18 19	17 53	17 38	17 20
12	23 18	22 37	22 07	21 44	21 25	21 08	20 41	20 17	19 54	19 32	19 08	18 41	18 24	18 05
13	.. ..	23 26	22 57	22 35	22 16	22 00	21 32	21 09	20 47	20 25	20 01	19 34	19 17	18 58
14	0 06	.. ..	23 39	23 18	23 01	22 47	22 21	22 00	21 39	21 19	20 57	20 31	20 16	19 58
15	0 39	0 04	.. ..	23 55	23 41	23 28	23 07	22 48	22 30	22 12	21 53	21 31	21 18	21 03
16	1 01	0 33	0 12	.. ..	.. ..	.. ..	23 48	23 33	23 19	23 05	22 50	22 32	22 22	22 10
17	1 16	0 55	0 40	0 26	0 15	0 05	.. ..	.. ..	.. ..	23 57	23 46	23 34	23 27	23 18

.. .. indicates phenomenon will occur the next day.

UNIVERSAL TIME FOR MERIDIAN OF GREENWICH

MOONRISE

Lat.	+40°	+42°	+44°	+46°	+48°	+50°	+52°	+54°	+56°	+58°	+60°	+62°	+64°	+66°
	h m	h m	h m	h m	h m	h m	h m	h m	h m	h m	h m	h m	h m	h m
Oct. 24	15 58	15 57	15 55	15 53	15 51	15 49	15 47	15 45	15 42	15 40	15 36	15 33	15 29	15 24
25	16 26	16 23	16 19	16 16	16 11	16 07	16 02	15 57	15 51	15 44	15 37	15 29	15 19	15 08
26	16 59	16 54	16 48	16 42	16 36	16 29	16 21	16 12	16 03	15 52	15 39	15 25	15 07	14 46
27	17 39	17 32	17 25	17 17	17 08	16 58	16 47	16 35	16 21	16 04	15 45	15 21	14 48	13 57
28	18 29	18 21	18 12	18 02	17 51	17 39	17 25	17 10	16 51	16 29	16 00	15 17	▭	▭
29	19 29	19 20	19 11	19 00	18 48	18 35	18 21	18 03	17 42	17 16	16 40	15 20	▭	▭
30	20 37	20 29	20 20	20 10	19 59	19 47	19 33	19 17	18 57	18 34	18 03	17 12	▭	▭
31	21 49	21 41	21 34	21 25	21 16	21 06	20 54	20 41	20 26	20 09	19 47	19 18	18 35	▭
Nov. 1	22 59	22 54	22 48	22 41	22 34	22 26	22 18	22 08	21 57	21 45	21 30	21 12	20 50	20 21
2	.. ..	.. ..	23 58	23 54	23 49	23 44	23 38	23 31	23 24	23 16	23 06	22 55	22 42	22 27
3	0 07	0 03	.. ..	.. ..	.. ..	.. ..	.. ..	.. ..	.. ..	.. ..	.. ..	.. ..	.. ..	.. ..
4	1 11	1 08	1 06	1 03	1 00	0 57	0 54	0 50	0 46	0 41	0 36	0 30	0 23	0 15
5	2 12	2 12	2 11	2 10	2 09	2 08	2 07	2 05	2 04	2 03	2 01	1 59	1 57	1 54
6	3 13	3 13	3 14	3 15	3 16	3 17	3 18	3 19	3 20	3 22	3 23	3 25	3 27	3 30
7	4 12	4 15	4 17	4 20	4 22	4 25	4 29	4 32	4 36	4 41	4 46	4 52	4 58	5 06
8	5 13	5 17	5 21	5 25	5 30	5 35	5 40	5 46	5 53	6 01	6 10	6 20	6 32	6 47
9	6 14	6 19	6 25	6 31	6 38	6 45	6 53	7 02	7 12	7 24	7 37	7 53	8 13	8 39
10	7 16	7 23	7 30	7 38	7 47	7 56	8 07	8 18	8 32	8 48	9 08	9 32	10 06	11 06
11	8 18	8 26	8 34	8 44	8 54	9 05	9 18	9 33	9 51	10 12	10 39	11 18	■	■
12	9 17	9 26	9 35	9 45	9 57	10 10	10 24	10 41	11 01	11 27	12 02	13 13	■	■
13	10 11	10 20	10 30	10 40	10 52	11 05	11 19	11 37	11 57	12 23	13 00	14 27	■	■
14	10 59	11 07	11 16	11 26	11 36	11 48	12 02	12 18	12 36	12 58	13 27	14 11	■	■
15	11 39	11 46	11 54	12 02	12 12	12 22	12 33	12 46	13 01	13 18	13 39	14 06	14 43	16 12
16	12 13	12 19	12 25	12 32	12 39	12 47	12 56	13 05	13 16	13 29	13 44	14 01	14 23	14 50
17	12 42	12 46	12 51	12 56	13 01	13 07	13 13	13 20	13 27	13 36	13 46	13 57	14 10	14 26

MOONSET

Lat.	+40°	+42°	+44°	+46°	+48°	+50°	+52°	+54°	+56°	+58°	+60°	+62°	+64°	+66°
	h m	h m	h m	h m	h m	h m	h m	h m	h m	h m	h m	h m	h m	h m
Oct. 24	4 02	4 02	4 03	4 04	4 05	4 06	4 07	4 08	4 09	4 10	4 12	4 13	4 15	4 17
25	5 19	5 22	5 25	5 28	5 31	5 34	5 38	5 42	5 47	5 52	5 58	6 05	6 13	6 23
26	6 40	6 44	6 49	6 54	7 00	7 07	7 14	7 21	7 30	7 40	7 51	8 05	8 21	8 41
27	8 02	8 09	8 16	8 23	8 32	8 41	8 51	9 03	9 16	9 32	9 50	10 14	10 45	11 36
28	9 23	9 31	9 40	9 49	10 00	10 12	10 25	10 40	10 58	11 20	11 49	12 31	▭	▭
29	10 37	10 45	10 55	11 05	11 17	11 30	11 45	12 02	12 23	12 49	13 25	14 46	▭	▭
30	11 38	11 47	11 56	12 06	12 17	12 30	12 44	13 00	13 20	13 44	14 15	15 06	▭	▭
31	12 27	12 35	12 43	12 52	13 01	13 12	13 24	13 37	13 53	14 11	14 34	15 03	15 46	▭
Nov. 1	13 05	13 11	13 18	13 25	13 33	13 41	13 50	14 00	14 12	14 25	14 41	14 59	15 22	15 52
2	13 36	13 40	13 45	13 50	13 56	14 02	14 08	14 16	14 24	14 33	14 43	14 55	15 09	15 26
3	14 01	14 04	14 07	14 10	14 14	14 18	14 22	14 27	14 32	14 37	14 44	14 51	15 00	15 09
4	14 23	14 24	14 26	14 27	14 29	14 31	14 33	14 35	14 38	14 41	14 44	14 47	14 51	14 56
5	14 43	14 43	14 43	14 43	14 43	14 43	14 43	14 43	14 43	14 43	14 43	14 43	14 43	14 43
6	15 03	15 01	15 00	14 58	14 56	14 55	14 53	14 50	14 48	14 45	14 42	14 39	14 35	14 31
7	15 23	15 21	15 18	15 14	15 11	15 07	15 03	14 59	14 54	14 48	14 42	14 35	14 27	14 18
8	15 46	15 42	15 37	15 32	15 27	15 21	15 15	15 08	15 01	14 52	14 42	14 31	14 18	14 02
9	16 12	16 06	16 00	15 54	15 47	15 39	15 30	15 21	15 10	14 58	14 43	14 27	14 06	13 40
10	16 43	16 36	16 28	16 20	16 11	16 01	15 50	15 38	15 24	15 07	14 47	14 22	13 48	12 47
11	17 20	17 12	17 03	16 54	16 43	16 32	16 18	16 03	15 46	15 24	14 57	14 17	■	■
12	18 05	17 56	17 47	17 37	17 25	17 12	16 58	16 41	16 20	15 55	15 19	14 08	■	■
13	18 58	18 49	18 40	18 30	18 18	18 05	17 50	17 33	17 13	16 47	16 11	14 44	■	■
14	19 58	19 50	19 41	19 32	19 21	19 09	18 56	18 41	18 23	18 01	17 32	16 49	■	■
15	21 03	20 56	20 48	20 40	20 32	20 22	20 11	19 59	19 44	19 27	19 07	18 41	18 04	16 36
16	22 10	22 05	21 59	21 53	21 46	21 38	21 30	21 21	21 11	20 59	20 45	20 28	20 08	19 41
17	23 18	23 15	23 11	23 06	23 02	22 57	22 51	22 45	22 38	22 31	22 22	22 12	22 00	21 46

▭ indicates Moon continuously above horizon.
■ indicates Moon continuously below horizon.
.. .. indicates phenomenon will occur the next day.

UNIVERSAL TIME FOR MERIDIAN OF GREENWICH

MOONRISE

Lat.	−55°	−50°	−45°	−40°	−35°	−30°	−20°	−10°	0°	+10°	+20°	+30°	+35°	+40°
	h m	h m	h m	h m	h m	h m	h m	h m	h m	h m	h m	h m	h m	h m
Nov. 16	8 36	9 02	9 21	9 37	9 51	10 03	10 23	10 40	10 56	11 12	11 29	11 49	12 00	12 13
17	10 01	10 19	10 34	10 45	10 55	11 04	11 19	11 32	11 44	11 56	12 09	12 24	12 32	12 42
18	11 27	11 38	11 47	11 54	12 01	12 06	12 15	12 24	12 31	12 39	12 47	12 56	13 02	13 08
19	12 54	12 58	13 02	13 04	13 07	13 09	13 12	13 15	13 18	13 21	13 24	13 28	13 30	13 32
20	14 24	14 21	14 18	14 16	14 14	14 13	14 10	14 08	14 06	14 04	14 02	13 59	13 58	13 56
21	15 57	15 46	15 38	15 31	15 25	15 20	15 11	15 03	14 56	14 49	14 41	14 33	14 28	14 22
22	17 36	17 17	17 02	16 50	16 40	16 31	16 15	16 02	15 50	15 37	15 25	15 10	15 01	14 52
23	19 19	18 51	18 29	18 12	17 57	17 45	17 24	17 05	16 48	16 31	16 14	15 53	15 41	15 28
24	21 00	20 22	19 55	19 33	19 16	19 00	18 34	18 12	17 52	17 31	17 09	16 44	16 30	16 13
25	22 25	21 42	21 12	20 49	20 29	20 13	19 44	19 20	18 58	18 36	18 12	17 44	17 28	17 09
26	23 24	22 44	22 15	21 52	21 33	21 17	20 50	20 26	20 04	19 42	19 18	18 51	18 34	18 16
27	23 59	23 26	23 02	22 42	22 26	22 11	21 47	21 26	21 06	20 46	20 25	20 00	19 46	19 29
28	.. ..	23 55	23 36	23 21	23 07	22 56	22 36	22 19	22 02	21 46	21 29	21 08	20 56	20 43
29	0 20	.. ..	.. ..	23 51	23 41	23 33	23 18	23 05	22 53	22 41	22 28	22 13	22 04	21 54
30	0 33	0 16	0 02	.. ..	.. ..	.. ..	23 55	23 47	23 39	23 31	23 22	23 13	23 07	23 01
Dec. 1	0 42	0 32	0 23	0 16	0 10	0 05	.. ..	.. ..	.. ..	.. ..	.. ..	.. ..	.. ..	.. ..
2	0 50	0 45	0 41	0 38	0 35	0 33	0 29	0 25	0 21	0 18	0 14	0 10	0 07	0 04
3	0 56	0 57	0 58	0 58	0 59	0 59	1 00	1 01	1 02	1 02	1 03	1 04	1 05	1 05
4	1 03	1 09	1 14	1 19	1 22	1 26	1 31	1 37	1 42	1 46	1 52	1 58	2 01	2 05
5	1 10	1 22	1 32	1 40	1 47	1 53	2 03	2 13	2 22	2 31	2 41	2 52	2 58	3 05
6	1 19	1 37	1 51	2 03	2 13	2 22	2 37	2 51	3 04	3 17	3 30	3 46	3 56	4 06
7	1 31	1 55	2 14	2 30	2 43	2 55	3 14	3 32	3 48	4 04	4 22	4 42	4 54	5 08
8	1 49	2 20	2 43	3 02	3 18	3 32	3 55	4 16	4 35	4 55	5 15	5 39	5 54	6 10
9	2 15	2 52	3 20	3 41	3 59	4 15	4 41	5 04	5 25	5 47	6 10	6 36	6 52	7 11
10	2 55	3 36	4 05	4 28	4 47	5 04	5 31	5 55	6 17	6 39	7 03	7 31	7 48	8 07

MOONSET

Lat.	−55°	−50°	−45°	−40°	−35°	−30°	−20°	−10°	0°	+10°	+20°	+30°	+35°	+40°
	h m	h m	h m	h m	h m	h m	h m	h m	h m	h m	h m	h m	h m	h m
Nov. 16	1 01	0 33	0 12	.. ..	.. ..	.. ..	23 48	23 33	23 19	23 05	22 50	22 32	22 22	22 10
17	1 16	0 55	0 40	0 26	0 15	0 05	.. ..	.. ..	.. ..	23 57	23 46	23 34	23 27	23 18
18	1 27	1 13	1 03	0 54	0 46	0 39	0 27	0 17	0 07	.. ..	.. ..	.. ..	.. ..	.. ..
19	1 36	1 29	1 23	1 18	1 14	1 11	1 05	0 59	0 54	0 48	0 42	0 36	0 32	0 28
20	1 44	1 43	1 43	1 42	1 42	1 42	1 41	1 41	1 41	1 40	1 40	1 39	1 39	1 38
21	1 52	1 58	2 03	2 07	2 11	2 14	2 20	2 24	2 29	2 34	2 39	2 44	2 48	2 51
22	2 02	2 15	2 26	2 35	2 42	2 49	3 01	3 11	3 21	3 31	3 41	3 53	4 00	4 08
23	2 15	2 36	2 53	3 07	3 18	3 29	3 47	4 02	4 17	4 31	4 47	5 06	5 16	5 28
24	2 35	3 05	3 28	3 47	4 02	4 16	4 39	4 59	5 18	5 37	5 57	6 21	6 35	6 51
25	3 08	3 47	4 15	4 37	4 55	5 11	5 38	6 01	6 23	6 45	7 08	7 35	7 51	8 10
26	4 02	4 45	5 15	5 39	5 58	6 15	6 43	7 07	7 30	7 52	8 16	8 44	9 01	9 20
27	5 20	5 59	6 27	6 50	7 08	7 24	7 50	8 13	8 34	8 55	9 18	9 44	9 59	10 17
28	6 51	7 22	7 45	8 04	8 20	8 33	8 56	9 15	9 34	9 52	10 11	10 33	10 46	11 01
29	8 23	8 45	9 03	9 17	9 29	9 40	9 57	10 13	10 27	10 41	10 57	11 14	11 24	11 35
30	9 50	10 05	10 17	10 26	10 35	10 42	10 54	11 05	11 15	11 25	11 36	11 48	11 55	12 03
Dec. 1	11 11	11 20	11 26	11 32	11 36	11 41	11 48	11 54	11 59	12 05	12 11	12 18	12 22	12 26
2	12 29	12 31	12 33	12 34	12 35	12 36	12 38	12 39	12 41	12 42	12 44	12 45	12 46	12 47
3	13 44	13 40	13 37	13 34	13 32	13 30	13 27	13 24	13 21	13 18	13 15	13 12	13 10	13 07
4	14 59	14 49	14 41	14 34	14 29	14 24	14 15	14 08	14 01	13 54	13 47	13 38	13 33	13 28
5	16 14	15 58	15 45	15 35	15 26	15 18	15 05	14 53	14 42	14 31	14 19	14 06	13 59	13 50
6	17 31	17 08	16 51	16 36	16 24	16 13	15 55	15 39	15 25	15 10	14 55	14 37	14 27	14 15
7	18 49	18 19	17 56	17 38	17 23	17 10	16 48	16 28	16 10	15 53	15 34	15 12	14 59	14 44
8	20 04	19 27	19 01	18 40	18 22	18 07	17 41	17 19	16 59	16 39	16 17	15 52	15 37	15 20
9	21 11	20 30	20 01	19 38	19 19	19 03	18 36	18 12	17 50	17 28	17 05	16 37	16 21	16 02
10	22 04	21 23	20 54	20 31	20 12	19 56	19 29	19 05	18 43	18 21	17 57	17 29	17 13	16 53

.. .. indicates phenomenon will occur the next day.

UNIVERSAL TIME FOR MERIDIAN OF GREENWICH

MOONRISE

Lat.	+40°	+42°	+44°	+46°	+48°	+50°	+52°	+54°	+56°	+58°	+60°	+62°	+64°	+66°
	h m	h m	h m	h m	h m	h m	h m	h m	h m	h m	h m	h m	h m	h m
Nov. 16	12 13	12 19	12 25	12 32	12 39	12 47	12 56	13 05	13 16	13 29	13 44	14 01	14 23	14 50
17	12 42	12 46	12 51	12 56	13 01	13 07	13 13	13 20	13 27	13 36	13 46	13 57	14 10	14 26
18	13 08	13 10	13 13	13 16	13 20	13 23	13 27	13 31	13 36	13 41	13 46	13 53	14 00	14 09
19	13 32	13 33	13 34	13 35	13 37	13 38	13 39	13 41	13 43	13 44	13 47	13 49	13 52	13 55
20	13 56	13 56	13 55	13 54	13 53	13 52	13 52	13 50	13 49	13 48	13 47	13 45	13 43	13 41
21	14 22	14 20	14 17	14 15	14 12	14 08	14 05	14 01	13 57	13 52	13 47	13 41	13 35	13 27
22	14 52	14 48	14 43	14 38	14 33	14 27	14 21	14 14	14 07	13 58	13 49	13 38	13 25	13 10
23	15 28	15 22	15 15	15 08	15 01	14 52	14 43	14 33	14 21	14 08	13 53	13 34	13 12	12 42
24	16 13	16 05	15 57	15 48	15 38	15 27	15 15	15 01	14 45	14 26	14 03	13 32	12 45	▭
25	17 09	17 00	16 51	16 41	16 29	16 17	16 02	15 46	15 26	15 02	14 29	13 34	▭	▭
26	18 16	18 07	17 58	17 47	17 36	17 23	17 09	16 52	16 32	16 07	15 33	14 31	▭	▭
27	19 29	19 21	19 13	19 04	18 54	18 43	18 30	18 16	17 59	17 39	17 14	16 38	15 30	▭
28	20 43	20 36	20 30	20 23	20 15	20 06	19 56	19 46	19 33	19 19	19 02	18 41	18 13	17 34
29	21 54	21 49	21 44	21 39	21 34	21 27	21 21	21 13	21 05	20 55	20 44	20 31	20 15	19 56
30	23 01	22 58	22 55	22 52	22 48	22 44	22 40	22 35	22 30	22 25	22 18	22 11	22 02	21 52
Dec. 1	.. ..	.. ..	.. ..	.. ..	23 59	23 57	23 55	23 53	23 51	23 49	23 46	23 43	23 39	23 35
2	0 04	0 03	0 02	0 00	.. ..	.. ..	.. ..	.. ..	.. ..	.. ..	.. ..	.. ..	.. ..	.. ..
3	1 05	1 06	1 06	1 06	1 07	1 07	1 07	1 08	1 08	1 09	1 10	1 10	1 11	1 12
4	2 05	2 07	2 09	2 11	2 13	2 16	2 18	2 21	2 24	2 28	2 32	2 37	2 42	2 48
5	3 05	3 09	3 12	3 16	3 20	3 25	3 29	3 35	3 41	3 48	3 55	4 04	4 14	4 27
6	4 06	4 11	4 16	4 22	4 28	4 34	4 42	4 50	4 59	5 09	5 21	5 35	5 52	6 13
7	5 08	5 14	5 21	5 28	5 36	5 45	5 55	6 06	6 18	6 33	6 50	7 12	7 40	8 21
8	6 10	6 18	6 26	6 35	6 44	6 55	7 07	7 21	7 38	7 57	8 21	8 55	9 53	■
9	7 11	7 19	7 28	7 38	7 49	8 02	8 16	8 32	8 51	9 16	9 48	10 44	■	■
10	8 07	8 16	8 25	8 35	8 47	9 00	9 15	9 32	9 53	10 19	10 55	12 24	■	■

MOONSET

Lat.	+40°	+42°	+44°	+46°	+48°	+50°	+52°	+54°	+56°	+58°	+60°	+62°	+64°	+66°
	h m	h m	h m	h m	h m	h m	h m	h m	h m	h m	h m	h m	h m	h m
Nov. 16	22 10	22 05	21 59	21 53	21 46	21 38	21 30	21 21	21 11	20 59	20 45	20 28	20 08	19 41
17	23 18	23 15	23 11	23 06	23 02	22 57	22 51	22 45	22 38	22 31	22 22	22 12	22 00	21 46
18	.. ..	.. ..	.. ..	.. ..	.. ..	.. ..	.. ..	.. ..	.. ..	.. ..	23 58	23 53	23 47	23 41
19	0 28	0 26	0 24	0 21	0 19	0 16	0 13	0 10	0 07	0 03	.. ..	.. ..	.. ..	.. ..
20	1 38	1 38	1 38	1 38	1 37	1 37	1 37	1 37	1 36	1 36	1 35	1 35	1 34	1 34
21	2 51	2 53	2 55	2 57	2 59	3 01	3 03	3 06	3 09	3 12	3 16	3 20	3 25	3 30
22	4 08	4 12	4 15	4 19	4 24	4 29	4 34	4 40	4 46	4 54	5 02	5 12	5 23	5 37
23	5 28	5 34	5 40	5 46	5 53	6 01	6 09	6 19	6 30	6 42	6 56	7 14	7 35	8 04
24	6 51	6 58	7 06	7 14	7 24	7 34	7 46	7 59	8 15	8 33	8 56	9 26	10 12	▭
25	8 10	8 18	8 28	8 38	8 49	9 01	9 15	9 32	9 51	10 16	10 48	11 43	▭	▭
26	9 20	9 28	9 38	9 48	10 00	10 13	10 27	10 44	11 04	11 29	12 04	13 05	▭	▭
27	10 17	10 24	10 33	10 42	10 53	11 04	11 17	11 32	11 49	12 09	12 35	13 11	14 20	▭
28	11 01	11 07	11 14	11 22	11 30	11 40	11 50	12 01	12 14	12 29	12 47	13 09	13 37	14 18
29	11 35	11 40	11 46	11 51	11 58	12 04	12 12	12 20	12 29	12 40	12 52	13 06	13 22	13 43
30	12 03	12 06	12 10	12 14	12 18	12 23	12 28	12 33	12 39	12 46	12 54	13 02	13 12	13 24
Dec. 1	12 26	12 28	12 30	12 32	12 35	12 37	12 40	12 43	12 46	12 50	12 54	12 59	13 04	13 10
2	12 47	12 48	12 48	12 49	12 49	12 50	12 51	12 51	12 52	12 53	12 54	12 55	12 56	12 57
3	13 07	13 06	13 05	13 04	13 03	13 02	13 00	12 59	12 57	12 55	12 53	12 51	12 48	12 45
4	13 28	13 26	13 23	13 20	13 17	13 14	13 11	13 07	13 03	12 58	12 53	12 47	12 40	12 33
5	13 50	13 46	13 42	13 38	13 33	13 28	13 22	13 16	13 09	13 02	12 53	12 43	12 32	12 18
6	14 15	14 10	14 04	13 58	13 52	13 44	13 37	13 28	13 18	13 07	12 54	12 39	12 21	11 59
7	14 44	14 38	14 31	14 23	14 15	14 05	13 55	13 44	13 31	13 15	12 57	12 35	12 07	11 25
8	15 20	15 12	15 03	14 54	14 44	14 33	14 21	14 06	13 50	13 30	13 05	12 32	11 33	■
9	16 02	15 54	15 45	15 34	15 23	15 11	14 57	14 40	14 20	13 56	13 23	12 28	■	■
10	16 53	16 45	16 35	16 25	16 13	16 00	15 46	15 28	15 08	14 42	14 06	12 37	■	■

▭ indicates Moon continuously above horizon.
■ indicates Moon continuously below horizon.
.. .. indicates phenomenon will occur the next day.

UNIVERSAL TIME FOR MERIDIAN OF GREENWICH

MOONRISE

Lat.	−55°	−50°	−45°	−40°	−35°	−30°	−20°	−10°	0°	+10°	+20°	+30°	+35°	+40°
	h m	h m	h m	h m	h m	h m	h m	h m	h m	h m	h m	h m	h m	h m
Dec. 9	2 15	2 52	3 20	3 41	3 59	4 15	4 41	5 04	5 25	5 47	6 10	6 36	6 52	7 11
10	2 55	3 36	4 05	4 28	4 47	5 04	5 31	5 55	6 17	6 39	7 03	7 31	7 48	8 07
11	3 52	4 32	5 01	5 23	5 42	5 58	6 25	6 48	7 10	7 32	7 55	8 22	8 38	8 57
12	5 04	5 39	6 04	6 25	6 42	6 56	7 21	7 43	8 03	8 22	8 44	9 08	9 22	9 39
13	6 24	6 52	7 13	7 30	7 45	7 57	8 18	8 36	8 53	9 10	9 28	9 49	10 01	10 15
14	7 49	8 09	8 25	8 38	8 49	8 58	9 14	9 29	9 42	9 55	10 09	10 25	10 34	10 45
15	9 13	9 26	9 37	9 45	9 52	9 59	10 10	10 19	10 28	10 37	10 47	10 58	11 04	11 11
16	10 38	10 44	10 49	10 53	10 56	11 00	11 05	11 10	11 14	11 18	11 23	11 28	11 31	11 35
17	12 03	12 02	12 02	12 01	12 01	12 01	12 00	12 00	12 00	11 59	11 59	11 59	11 58	11 58
18	13 31	13 23	13 17	13 12	13 08	13 04	12 58	12 52	12 47	12 41	12 36	12 30	12 26	12 22
19	15 03	14 48	14 36	14 26	14 18	14 10	13 58	13 47	13 37	13 26	13 16	13 04	12 57	12 49
20	16 41	16 17	15 59	15 44	15 31	15 20	15 02	14 46	14 31	14 16	14 00	13 42	13 32	13 20
21	18 21	17 48	17 23	17 03	16 47	16 33	16 09	15 49	15 30	15 11	14 51	14 28	14 15	13 59
22	19 54	19 13	18 44	18 21	18 02	17 46	17 19	16 56	16 34	16 12	15 49	15 23	15 07	14 49
23	21 06	20 24	19 54	19 31	19 11	18 55	18 27	18 03	17 40	17 18	16 54	16 26	16 09	15 50
24	21 53	21 16	20 49	20 28	20 10	19 55	19 29	19 06	18 45	18 24	18 01	17 35	17 19	17 01
25	22 20	21 52	21 31	21 13	20 58	20 45	20 23	20 04	19 46	19 28	19 08	18 46	18 32	18 17
26	22 37	22 17	22 01	21 48	21 37	21 27	21 10	20 55	20 41	20 27	20 11	19 54	19 44	19 32
27	22 49	22 36	22 25	22 16	22 08	22 02	21 50	21 40	21 30	21 20	21 10	20 58	20 51	20 43
28	22 57	22 50	22 45	22 40	22 36	22 32	22 26	22 20	22 15	22 10	22 04	21 58	21 54	21 50
29	23 04	23 03	23 02	23 01	23 01	23 00	22 59	22 58	22 57	22 57	22 56	22 55	22 54	22 54
30	23 11	23 15	23 19	23 22	23 25	23 27	23 31	23 35	23 38	23 42	23 45	23 50	23 52	23 55
31	23 18	23 28	23 36	23 43	23 49	23 54	.. ..	.. ..	.. ..	.. ..	.. ..	.. ..	.. ..	.. ..
32	23 26	23 42	23 55	.. ..	.. ..	.. ..	0 03	0 11	0 19	0 26	0 35	0 44	0 50	0 56
33	23 37	23 59	.. ..	0 05	0 14	0 23	0 36	0 49	1 00	1 12	1 24	1 39	1 47	1 57

MOONSET

Lat.	−55°	−50°	−45°	−40°	−35°	−30°	−20°	−10°	0°	+10°	+20°	+30°	+35°	+40°
	h m	h m	h m	h m	h m	h m	h m	h m	h m	h m	h m	h m	h m	h m
Dec. 9	21 11	20 30	20 01	19 38	19 19	19 03	18 36	18 12	17 50	17 28	17 05	16 37	16 21	16 02
10	22 04	21 23	20 54	20 31	20 12	19 56	19 29	19 05	18 43	18 21	17 57	17 29	17 13	16 53
11	22 41	22 05	21 38	21 17	21 00	20 45	20 19	19 56	19 36	19 15	18 52	18 26	18 10	17 52
12	23 05	22 36	22 14	21 56	21 41	21 28	21 05	20 46	20 27	20 09	19 49	19 25	19 12	18 56
13	23 22	23 00	22 43	22 29	22 17	22 06	21 48	21 32	21 17	21 02	20 45	20 26	20 15	20 02
14	23 34	23 19	23 07	22 57	22 48	22 40	22 27	22 15	22 04	21 53	21 41	21 27	21 19	21 10
15	23 43	23 35	23 28	23 22	23 17	23 12	23 04	22 57	22 50	22 43	22 36	22 28	22 23	22 18
16	23 51	23 49	23 47	23 45	23 43	23 42	23 40	23 38	23 36	23 34	23 31	23 29	23 27	23 25
17	23 59	.. ..	.. ..	.. ..	.. ..	.. ..	.. ..	.. ..	.. ..	.. ..	.. ..	.. ..	.. ..	.. ..
18	.. ..	0 03	0 06	0 08	0 11	0 13	0 16	0 19	0 22	0 24	0 27	0 31	0 33	0 35
19	0 08	0 18	0 26	0 33	0 39	0 45	0 54	1 02	1 10	1 17	1 26	1 35	1 41	1 47
20	0 19	0 37	0 50	1 02	1 12	1 21	1 36	1 49	2 02	2 14	2 28	2 43	2 52	3 03
21	0 35	1 01	1 20	1 37	1 51	2 03	2 23	2 41	2 58	3 15	3 34	3 55	4 07	4 21
22	1 00	1 34	2 00	2 20	2 38	2 52	3 18	3 40	4 00	4 20	4 43	5 08	5 23	5 41
23	1 42	2 23	2 53	3 16	3 35	3 51	4 19	4 43	5 06	5 28	5 52	6 19	6 36	6 55
24	2 48	3 30	3 59	4 23	4 42	4 58	5 26	5 50	6 12	6 34	6 57	7 24	7 40	7 59
25	4 15	4 50	5 16	5 37	5 54	6 09	6 34	6 55	7 15	7 34	7 55	8 20	8 34	8 50
26	5 50	6 16	6 37	6 53	7 07	7 19	7 39	7 56	8 13	8 29	8 46	9 05	9 17	9 30
27	7 22	7 40	7 55	8 06	8 16	8 25	8 40	8 53	9 05	9 17	9 29	9 44	9 52	10 01
28	8 49	9 00	9 08	9 16	9 22	9 27	9 36	9 45	9 52	9 59	10 07	10 16	10 21	10 27
29	10 10	10 15	10 18	10 21	10 23	10 26	10 29	10 33	10 36	10 39	10 42	10 45	10 47	10 50
30	11 28	11 26	11 25	11 23	11 22	11 21	11 20	11 18	11 17	11 16	11 14	11 13	11 12	11 11
31	12 44	12 36	12 30	12 24	12 20	12 16	12 09	12 03	11 58	11 52	11 46	11 39	11 36	11 31
32	14 00	13 46	13 34	13 25	13 17	13 10	12 59	12 48	12 39	12 29	12 19	12 07	12 00	11 53
33	15 16	14 56	14 40	14 26	14 15	14 05	13 49	13 34	13 21	13 07	12 53	12 37	12 27	12 17

.. .. indicates phenomenon will occur the next day.

UNIVERSAL TIME FOR MERIDIAN OF GREENWICH

MOONRISE

Lat.	+40°	+42°	+44°	+46°	+48°	+50°	+52°	+54°	+56°	+58°	+60°	+62°	+64°	+66°
	h m	h m	h m	h m	h m	h m	h m	h m	h m	h m	h m	h m	h m	h m
Dec. 9	7 11	7 19	7 28	7 38	7 49	8 02	8 16	8 32	8 51	9 16	9 48	10 44	■	■
10	8 07	8 16	8 25	8 35	8 47	9 00	9 15	9 32	9 53	10 19	10 55	12 24	■	■
11	8 57	9 05	9 14	9 24	9 35	9 47	10 01	10 17	10 36	11 00	11 31	12 20	■	■
12	9 39	9 47	9 55	10 03	10 13	10 24	10 36	10 49	11 05	11 23	11 46	12 16	13 01	■
13	10 15	10 21	10 27	10 35	10 42	10 51	11 00	11 11	11 23	11 36	11 53	12 12	12 36	13 09
14	10 45	10 49	10 54	11 00	11 05	11 12	11 19	11 26	11 35	11 44	11 55	12 08	12 23	12 41
15	11 11	11 14	11 17	11 21	11 25	11 29	11 33	11 38	11 44	11 50	11 56	12 04	12 13	12 24
16	11 35	11 36	11 38	11 40	11 42	11 44	11 46	11 48	11 51	11 54	11 57	12 00	12 04	12 09
17	11 58	11 58	11 58	11 58	11 58	11 58	11 57	11 57	11 57	11 57	11 57	11 57	11 56	11 56
18	12 22	12 20	12 19	12 17	12 14	12 12	12 10	12 07	12 04	12 01	11 57	11 53	11 48	11 43
19	12 49	12 45	12 42	12 38	12 34	12 29	12 24	12 18	12 12	12 06	11 58	11 49	11 39	11 28
20	13 20	13 15	13 10	13 04	12 57	12 50	12 42	12 34	12 24	12 13	12 01	11 46	11 29	11 07
21	13 59	13 53	13 45	13 37	13 28	13 19	13 08	12 56	12 42	12 26	12 07	11 43	11 12	10 23
22	14 49	14 41	14 32	14 22	14 11	14 00	13 46	13 31	13 13	12 51	12 23	11 43	▭	▭
23	15 50	15 41	15 32	15 22	15 10	14 57	14 43	14 26	14 05	13 40	13 05	11 59	▭	▭
24	17 01	16 53	16 44	16 34	16 24	16 12	15 58	15 42	15 24	15 01	14 31	13 44	▭	▭
25	18 17	18 10	18 03	17 54	17 46	17 36	17 25	17 12	16 58	16 41	16 20	15 53	15 15	▭
26	19 32	19 27	19 21	19 15	19 08	19 01	18 53	18 44	18 34	18 22	18 08	17 52	17 32	17 07
27	20 43	20 40	20 36	20 32	20 27	20 22	20 17	20 11	20 05	19 58	19 49	19 40	19 28	19 15
28	21 50	21 48	21 46	21 44	21 42	21 39	21 37	21 34	21 30	21 26	21 22	21 18	21 12	21 06
29	22 54	22 53	22 53	22 53	22 52	22 52	22 52	22 51	22 51	22 50	22 50	22 49	22 48	22 47
30	23 55	23 56	23 58	23 59	.. ..	.. ..	.. ..	.. ..	.. ..	.. ..	.. ..	.. ..	.. ..	.. ..
31	.. ..	.. ..	.. ..	.. ..	0 01	0 02	0 04	0 06	0 09	0 11	0 14	0 17	0 21	0 25
32	0 56	0 59	1 02	1 05	1 08	1 12	1 16	1 21	1 26	1 31	1 38	1 45	1 54	2 04
33	1 57	2 01	2 06	2 11	2 16	2 22	2 28	2 35	2 44	2 53	3 03	3 15	3 30	3 48

MOONSET

Lat.	+40°	+42°	+44°	+46°	+48°	+50°	+52°	+54°	+56°	+58°	+60°	+62°	+64°	+66°
	h m	h m	h m	h m	h m	h m	h m	h m	h m	h m	h m	h m	h m	h m
Dec. 9	16 02	15 54	15 45	15 34	15 23	15 11	14 57	14 40	14 20	13 56	13 23	12 28	■	■
10	16 53	16 45	16 35	16 25	16 13	16 00	15 46	15 28	15 08	14 42	14 06	12 37	■	■
11	17 52	17 44	17 35	17 25	17 14	17 02	16 48	16 33	16 14	15 51	15 20	14 31	■	■
12	18 56	18 49	18 41	18 33	18 23	18 13	18 01	17 48	17 33	17 15	16 53	16 23	15 39	■
13	20 02	19 57	19 51	19 44	19 37	19 29	19 20	19 10	18 58	18 45	18 30	18 11	17 48	17 16
14	21 10	21 06	21 01	20 57	20 51	20 46	20 40	20 33	20 25	20 16	20 06	19 55	19 41	19 24
15	22 18	22 15	22 12	22 10	22 07	22 03	22 00	21 56	21 51	21 46	21 41	21 34	21 27	21 18
16	23 25	23 25	23 24	23 23	23 22	23 21	23 20	23 19	23 17	23 16	23 14	23 12	23 10	23 07
17	.. ..	.. ..	.. ..	.. ..	.. ..	.. ..	.. ..	.. ..	.. ..	.. ..	.. ..	.. ..	.. ..	.. ..
18	0 35	0 36	0 37	0 38	0 39	0 40	0 42	0 43	0 45	0 47	0 49	0 52	0 55	0 58
19	1 47	1 50	1 53	1 56	1 59	2 03	2 07	2 12	2 17	2 22	2 29	2 36	2 44	2 54
20	3 03	3 07	3 12	3 17	3 23	3 30	3 37	3 45	3 53	4 03	4 15	4 28	4 45	5 05
21	4 21	4 28	4 35	4 42	4 51	5 00	5 10	5 21	5 35	5 50	6 08	6 31	7 02	7 50
22	5 41	5 49	5 57	6 07	6 17	6 29	6 42	6 57	7 15	7 36	8 04	8 43	▭	▭
23	6 55	7 04	7 13	7 24	7 35	7 48	8 03	8 20	8 40	9 05	9 40	10 46	▭	▭
24	7 59	8 07	8 16	8 26	8 37	8 50	9 04	9 20	9 39	10 02	10 32	11 19	▭	▭
25	8 50	8 57	9 05	9 14	9 23	9 33	9 45	9 58	10 13	10 31	10 52	11 19	11 58	▭
26	9 30	9 35	9 42	9 48	9 56	10 03	10 12	10 22	10 33	10 45	11 00	11 17	11 38	12 04
27	10 01	10 05	10 10	10 14	10 19	10 25	10 31	10 38	10 45	10 53	11 03	11 14	11 26	11 41
28	10 27	10 30	10 32	10 35	10 38	10 42	10 45	10 49	10 54	10 58	11 04	11 10	11 17	11 25
29	10 50	10 51	10 52	10 53	10 54	10 55	10 57	10 58	11 00	11 02	11 04	11 06	11 09	11 12
30	11 11	11 10	11 10	11 09	11 08	11 08	11 07	11 06	11 06	11 05	11 04	11 03	11 01	11 00
31	11 31	11 29	11 27	11 25	11 23	11 20	11 17	11 14	11 11	11 07	11 03	10 59	10 53	10 47
32	11 53	11 49	11 46	11 42	11 38	11 33	11 29	11 23	11 17	11 11	11 03	10 55	10 45	10 34
33	12 17	12 12	12 07	12 01	11 55	11 49	11 42	11 34	11 25	11 15	11 04	10 51	10 35	10 16

▭ indicates Moon continuously above horizon.
■ indicates Moon continuously below horizon.
.. .. indicates phenomenon will occur the next day.

There are four eclipses, two of the Sun and two of the Moon.

I	March 3–4	Total eclipse of the Moon
II	March 19	Partial eclipse of the Sun
III	August 28	Total eclipse of the Moon
IV	September 11	Partial eclipse of the Sun

Standard corrections of $+0\overset{''}{.}5$ and $-0\overset{''}{.}25$ have been applied to the longitude and latitude of the Moon, respectively, to help correct for the difference between the center of figure and the center of mass.

All time arguments are given provisionally in Universal Time, using $\Delta T(\mathrm{A}) = +65\overset{s}{.}0$. Once an updated value of ΔT is known, the data on these pages may be expressed in Universal Time as follows:

• Define $\delta T = \Delta T - \Delta T(\mathrm{A})$, in units of seconds of time.

• Change the times of circumstances given in preliminary Universal Time by subtracting δT.

• Correct the tabulated longitudes, $\lambda(\mathrm{A})$, according to the formula $\lambda = \lambda(\mathrm{A}) + 0.00417807\,\delta T$, where the longitudes have units of degrees.

• Leave all other quantities unchanged.

• The correction for δT is included in the Besselian elements.

Longitude is positive to the east, and negative to the west.

Explanation of Lunar Eclipse Diagrams

Information on lunar eclipses is presented in the form of a diagram consisting of two parts. The upper panel shows the path of the Moon relative to the penumbral and umbral shadows of the Earth. The lower panel shows the visibility of the eclipse from the surface of the Earth.

The title of the upper panel includes the type of eclipse, its place in the sequence of eclipses for the year and the Greenwich calendar date of the eclipse. The inner darker circle is the umbral shadow of the Earth and the outer lighter circle is that of the penumbra. The axis of the shadow of the Earth is denoted by (+) with the ecliptic shown for reference purposes. A 30-arcminute scale bar is provided on the right hand side of the diagram and the orientation is given by the cardinal points displayed on the small graphic on the left hand side of the diagram. The position angle (PA) is measured from North point of the lunar disk along the limb of the Moon to the point of contact. It is shown on the graphic by the use of an arc extending anti-clockwise (eastwards) from North terminated with an arrow head.

Moon symbols are plotted at the principal phases of the eclipse to show its position relative to the umbral and penumbral shadows. The UT times of the different phases of the eclipse to the nearest tenth of a minute are printed above or below the Moon symbols as appropriate. P1 and P4 are the first and last external contacts of the penumbra respectively and denote the beginning and end of the penumbral eclipse respectively. U1 and U4 are the first and last external contacts of the umbra denoting the beginning and end of the partial phase of the eclipse respectively. U2 and U3 are the first and last internal contacts of the umbra and denote the beginning and end of the total phase respectively. MID is the middle of the eclipse. The position angle is given for P1 and P4 for penumbral eclipses and U1 and U4 for partial and total eclipses. The UT time of the geocentric opposition in right ascension of the Sun and Moon and the magnitude of the eclipse are given above or below the Moon symbols as appropriate.

The lower panel is an cylindrical equidistant map projection showing the Earth centered on the longitude at which the Moon is in the zenith at the middle of the eclipse. The visibility of the eclipse is displayed by plotting the Moon rise/set terminator for the principal phases of the eclipse for which timing information is provided in the upper panel. The terminator for the middle of the eclipse is not plotted for the sake of clarity.

The unshaded area indicates the region of the Earth from which all the eclipse is visible whereas the darkest shading indicates the area from which the eclipse is invisible. The different shades of grey indicate regions where the Moon is either rising or setting during the principal phases of the eclipse. The Moon is rising on the left hand side of the diagram after the eclipse has started and is setting on the right hand side of the diagram before the eclipse ends. Labels are provided to this effect.

Symbols are plotted showing the locations for which the Moon is in the zenith at the principal phases of the eclipse. The points at which the Moon is in the zenith at P1 and P4 are denoted by (+), at U1 and U4 by ($\odot$) and at U2 and U3 by ($\oplus$). These symbols are also plotted on the upper panel where appropriate. The value of ΔT used for the calculation of the eclipse circumstances is given below the diagram. Country boundaries are also provided to assist the user in determining the visibility of the eclipse at a particular location.

Explanation of Solar Eclipse Diagrams

The solar eclipse diagrams show the region over which different phases of each eclipse may be seen and the times at which these phases occur. Each diagram has a series of dashed curves that show the outline of the Moon's penumbra on the Earth's surface at one-hour intervals. Short dashes show the leading edge and long dashes show the trailing edge. Except for certain extreme cases, the shadow outline moves generally from west to east. The Moon's shadow cone first contacts the Earth's surface where "First Contact" is indicated on the diagram. "Last Contact" is where the Moon's shadow cone last contacts the Earth's surface. The path of central eclipse, whether for a total, annular, or annular-total eclipse, is marked by two closely spaced curves that cut across all of the dashed curves. These two curves mark the extent of the Moon's umbral shadow on the Earth's surface. Viewers within these boundaries will observe a total, annular, or annular-total eclipse and viewers outside these boundaries will see a partial eclipse.

Solid curves labeled "Northern" and "Southern Limit of Eclipse" represent the furthest extent north or south of the Moon's penumbra on the Earth's surface. Viewers outside of these boundaries will not experience any eclipse. When only one of these two curves appears, only part of the Moon's penumbra touches the Earth; the other part is projected into space north or south of the Earth, and the terminator defines the other limit.

Another set of solid curves appears on some diagrams as two teardrop shapes (or lobes) on either end of the eclipse path, and on other diagrams as a distorted figure eight. These lobes represent in time the intersection of the Moon's penumbra with the Earth's terminator as the eclipse progresses. As time elapses, the Earth's terminator moves east-to-west while the Moon's penumbra moves west-to-east. These lobes connect to form an elongated figure eight on a diagram when part of the Moon's penumbra stays in contact with the Earth's terminator throughout the eclipse. The lobes become two separate teardrop shapes when the Moon's penumbra breaks contact with the Earth's terminator during the beginning of the eclipse and reconnects with it near the end. In the east, the outer portion of the lobe is labeled "Eclipse begins at Sunset" and marks the first contact between the Moon's penumbra and Earth's terminator in the east. Observers on this curve just fail to see the eclipse. The inner part of the lobe is labeled "Eclipse ends at Sunset" and marks the last contact between the Moon's penumbra and the Earth's terminator in the east. Observers on this curve just see the whole eclipse. The curve bisecting this lobe is labeled "Maximum Eclipse at Sunset" and is part of the sunset terminator at maximum eclipse. Viewers in the eastern half of the lobe will see the Sun set before maximum eclipse; *i.e.* see less than half of the eclipse. Viewers in the western half of the lobe will see the Sun set after maximum eclipse; *i.e.* see more than half of the eclipse. A similar description holds for the western lobe except everything occurs at sunrise instead of sunset.

I. - Total Eclipse of the Moon — 2007 March 03-04

UT of geocentric opposition in RA: March $3^d\ 23^h\ 0^m\ 43^s.959$

Umbral magnitude of the eclipse: 1.237

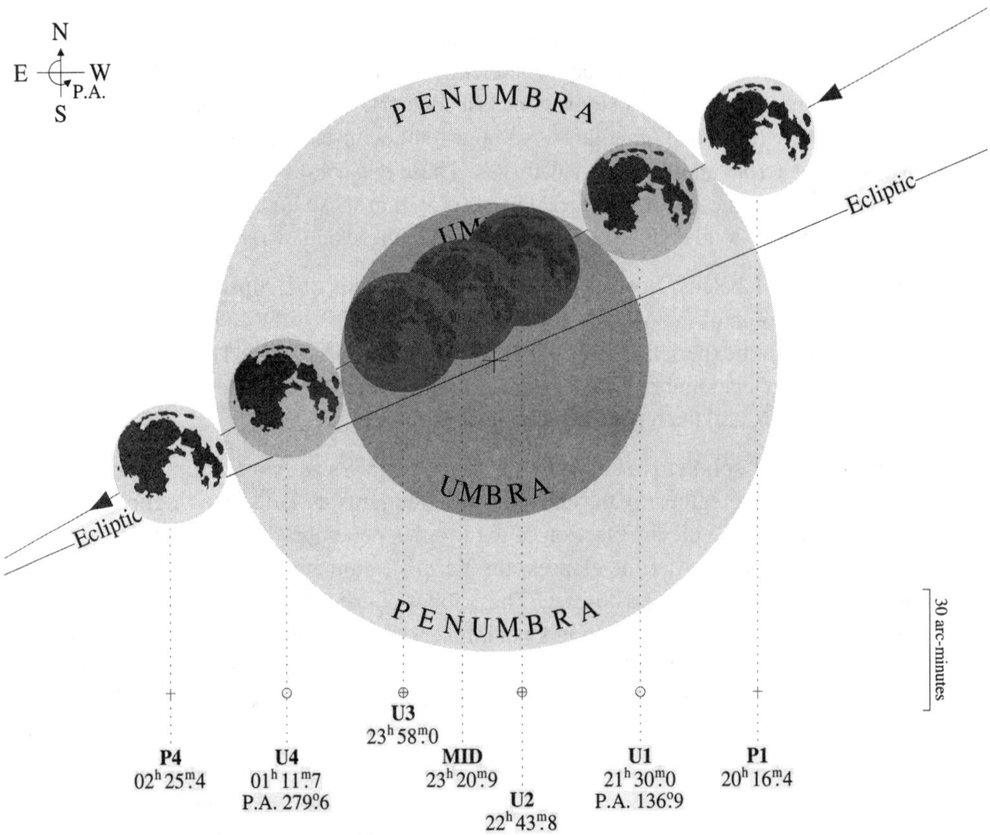

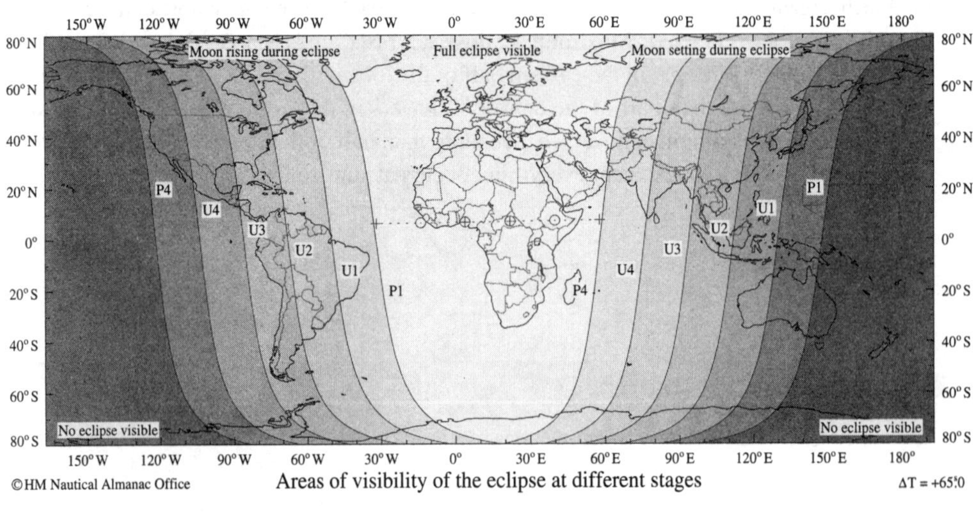

Areas of visibility of the eclipse at different stages

$\Delta T = +65^s.0$

II. –Partial Eclipse of the Sun, 2007 March 19

CIRCUMSTANCES OF THE ECLIPSE

UT of geocentric conjunction in right ascension, March $19^d\ 3^h\ 33^m\ 5\overset{s}{.}560$

Julian Date = 2454178.6479810190

			UT	Longitude	Latitude
		d	h m	° ′	° ′
Eclipse begins	March	19	0 38.3	+ 82 38.4	+15 26.3
Greatest eclipse		19	2 31.9	+ 55 24.4	+61 12.7
Eclipse ends		19	4 24.9	−156 40.7	+73 25.5

Magnitude of greatest eclipse: 0.876

BESSELIAN ELEMENTS

Let $t = (\mathrm{UT}-0^h) + \delta T/3600$ in units of hours.

These equations are valid over the range $0\overset{h}{.}542 \leq t \leq 4\overset{h}{.}583$. Do not use t outside the given range, and do not omit any terms in the series.

Intersection of axis of shadow with fundamental plane:

$$x = -1.80616707 + 0.50831854\,t + 0.00009758\,t^2 - 0.00000847\,t^3$$
$$y = +0.22656379 + 0.28142552\,t - 0.00000974\,t^2 - 0.00000490\,t^3$$

Direction of axis of shadow:

$$\sin d = -0.01385061 + 0.00027562\,t + 0.00000001\,t^2$$
$$\cos d = +0.99990397 + 0.00000398\,t - 0.00000011\,t^2 + 0.00000001\,t^3$$
$$\mu = 177\overset{\circ}{.}98923658 + 15.00427001\,t + 0.00000109\,t^2 - 0.00000005\,t^3 - 0.00417807\,\delta T$$

Radius of shadow on fundamental plane:

$$\text{penumbra} = +0.53629479 + 0.00002248\,t - 0.00001292\,t^2$$

$$\tan f_1 = +0.004697$$
$$\mu' = +0.261874 \text{ radians per hour}$$
$$d' = +0.000276 \text{ radians per hour}$$

III. - Total Eclipse of the Moon **2007 August 28**

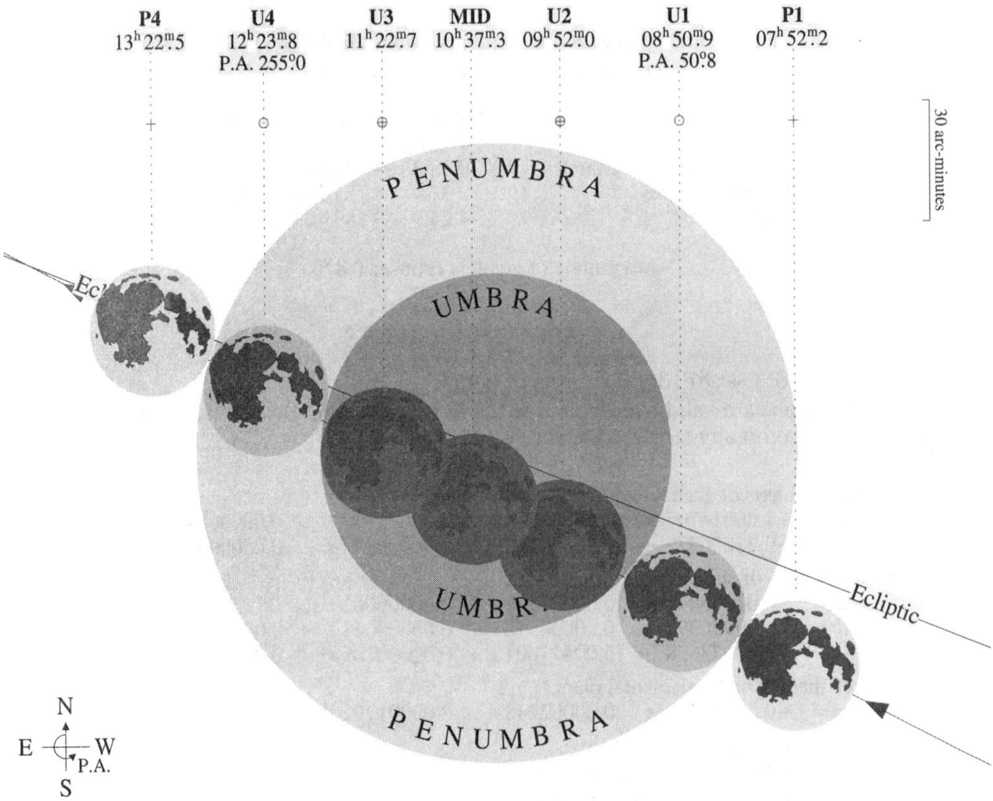

UT of geocentric opposition in RA: August 28^d 10^h 25^m $39^s.739$ Umbral magnitude of the eclipse: 1.481

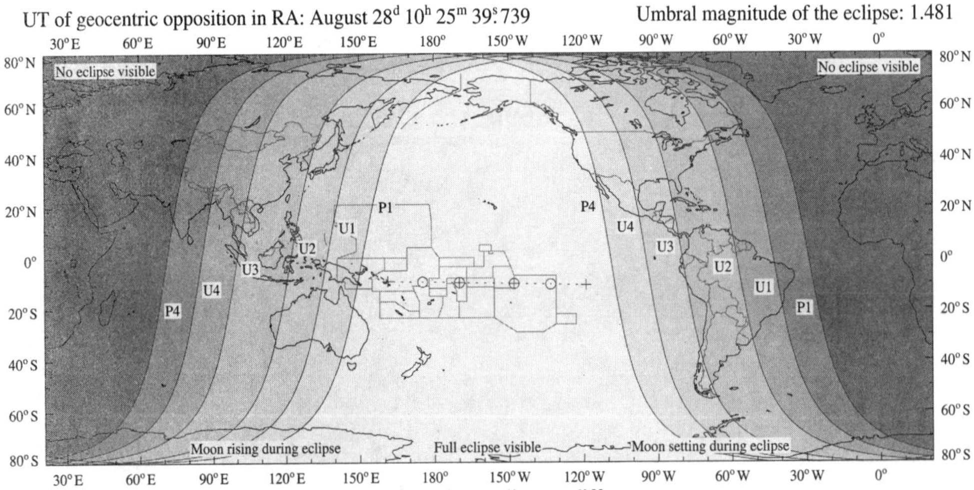

Areas of visibility of the eclipse at different stages

ΔT = +65s.0

IV. –Partial Eclipse of the Sun, 2007 September 11

CIRCUMSTANCES OF THE ECLIPSE

UT of geocentric conjunction in right ascension, September 11^{d} 13^{h} 42^{m} $40\overset{s}{.}421$

Julian Date = 2454355.0713011710

		UT		Longitude	Latitude
		d	h m	° ′	° ′
Eclipse begins	September	11	10 25.7	−65 46.3	−17 29.7
Greatest eclipse		11	12 31.3	−90 15.5	−61 09.8
Eclipse ends		11	14 36.5	+33 34.5	−74 17.8

Magnitude of greatest eclipse: 0.751

BESSELIAN ELEMENTS

Let $t = (\mathrm{UT} - 10^{h}) + \delta T / 3600$ in units of hours.

These equations are valid over the range $0\overset{h}{.}375 \le t \le 4\overset{h}{.}783$. Do not use t outside the given range, and do not omit any terms in the series. If μ is greater than 360°, then subtract 360° from its computed value.

Intersection of axis of shadow with fundamental plane:

$$x = -1.68440591 + 0.45380021\,t + 0.00003800\,t^2 - 0.00000535\,t^3$$
$$y = -0.36191403 - 0.24816025\,t + 0.00000011\,t^2 + 0.00000309\,t^3$$

Direction of axis of shadow:

$$\sin d = +0.08068326 - 0.00026506\,t - 0.00000002\,t^2$$
$$\cos d = +0.99673981 + 0.00002144\,t - 0.00000003\,t^2$$
$$\mu = 330\overset{\circ}{.}80312122 + 15.00474890\,t + 0.00000045\,t^2 - 0.00417807\,\delta T$$

Radius of shadow on fundamental plane:

$$\text{penumbra} = +0.56207242 + 0.00014368\,t - 0.00001017\,t^2$$

$$\tan f_1 = +0.004645$$
$$\mu' = +0.261882 \text{ radians per hour}$$
$$d' = -0.000266 \text{ radians per hour}$$

PARTIAL SOLAR ECLIPSE OF 2007 MARCH 19

PARTIAL SOLAR ECLIPSE OF 2007 SEPTEMBER 11

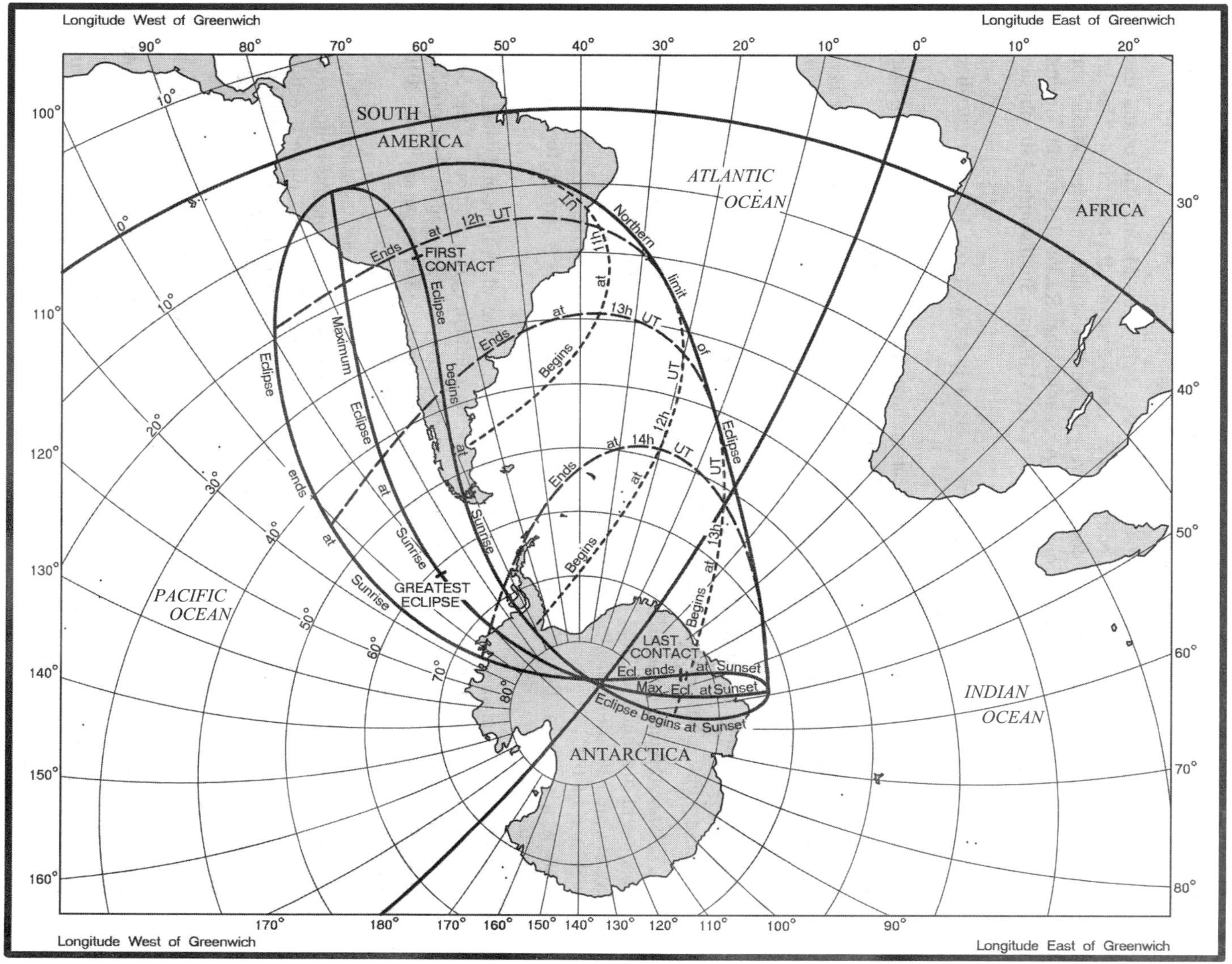

Joint publications of the Rutherford Appleton Laboratory and the United States Naval Observatory

These publications are published by and available from, The Stationery Office (UK), and the Superintendent of Documents, U.S. Government Printing Office except where noted. Their addresses are listed on the reverse of the title page of this volume.

The Astronomical Almanac contains ephemerides of the Sun, Moon, planets and their natural satellites, as well as data on eclipses and other astronomical phenomena. The data in this annual volume are calculated cooperatively by the British and American offices, with contributions from the Bureau des Longitudes, Astronomisches Rechen-Institut, the Institute of Theoretical Astronomy, St. Petersburg, and B.G. Marsden, Smithsonian Astrophysical Observatory.

The Nautical Almanac contains ephemerides at an interval of one hour and auxiliary astronomical data for marine navigation.

The Air Almanac contains ephemerides at an interval of ten minutes and auxiliary astronomical data for air navigation. It is not published by The Stationery Office.

Explanatory Supplement to The Astronomical Ephemeris and The American Ephemeris and Nautical Almanac is out of print. The new *Explanatory Supplement to The Astronomical Almanac* is available, see below.

Other publications of Rutherford Appleton Laboratory

These publications are available from The Stationery Office.

The Star Almanac for Land Surveyors contains the Greenwich hour angle of Aries and the position of the Sun, tabulated for every six hours, and represented by monthly polynomial coefficients. Positions of all stars brighter than magnitude 4·0 are tabulated monthly to a precision of $0^{s}\!.1$ in right ascension and $1''$ in declination. A CD-ROM accompanies this book which contains the electronic edition plus coefficients, in ASCII format, representing the data.

NavPac and Compact Data for 2006–2010 contains software, algorithms and data, which are mainly in the form of polynomial coefficients, for calculating the positions of the Sun, Moon, navigational planets and bright stars. It enables navigators to compute their position at sea from sextant observations using an IBM PC or compatible for the period 1986–2010. The tabular data are also supplied as ASCII files on the CD-ROM.

Planetary and Lunar Coordinates, 2001–2020 provides low-precision astronomical data and phenomena for use well in advance of the annual ephemerides. It contains heliocentric, geocentric, spherical and rectangular coordinates of the Sun, Moon and planets, eclipse maps and auxiliary data. All the tabular ephemerides are supplied solely on CD-ROM as ASCII and Adobe's portable document format files. The full printed edition is published in the United States by Willmann-Bell Inc.

Rapid Sight Reduction Tables for Navigation (AP3270 / NP 303), 3 volumes, formerly entitled *Sight Reduction Tables for Air Navigation.* Volume 1, selected stars for epoch 2005·0, containing the altitude to $1'$ and true azimuth to 1° for the seven stars most suitable for navigation, for all latitudes and hour angles of Aries. Volumes 2 and 3 contain altitudes to $1'$ and azimuths to 1° for integral degrees of declination from N 29° to S 29°, for relevant latitudes and all hour angles at which the zenith distance is less than 95° providing for sights of the Sun, Moon and planets.

Sight Reduction Tables for Marine Navigation (NP 401), 6 volumes. This series is designed to effect all solutions of the navigational triangle and is intended for use with *The Nautical Almanac.*

The UK Air Almanac contains data useful in the planning of activities where the level of illumination is important, particularly aircraft movements, and is produced to the general requirements of the Royal Air Force.

Other publications of the United States Naval Observatory

Astronomical Papers of the American Ephemeris† are issued irregularly and contain reports of research in celestial mechanics with particular relevance to ephemerides.

U.S. Naval Observatory Circulars† are issued irregularly to disseminate astronomical data concerning ephemerides or astronomical phenomena.

Explanatory Supplement to The Astronomical Almanac edited by P. Kenneth Seidelmann of the U.S. Naval Observatory. This book is an authoritative source on the basis and derivation of information contained in *The Astronomical Almanac*, and it contains material that is relevant to positional and dynamical astronomy and to chronology. It includes details of the FK5 J2000·0 reference system and transformations. The publication is a collaborative work with authors from the U.S. Naval Observatory, H.M. Nautical Almanac Office, the Jet Propulsion Laboratory and the Bureau des Longitudes. It is published by, and available from, University Science Books, 55D Gate Five Road, Sausalito, CA 94965, whose UK distributor is Macmillan.

MICA is an interactive astronomical almanac for professional applications. Software for both PC systems with Intel processors and Apple Macintosh computers is provided on a single CD-ROM. *MICA* allows a user to compute, to full precision, much of the tabular data contained in *The Astronomical Almanac*, as well as data for specific times. All calculations are made in real time and data are not interpolated from tables. MICA is published by, and available from, Willman-Bell Inc.

† These publications are available from the Nautical Almanac Office, U.S. Naval Observatory, Washington, DC 20392-5420.

Publications of other countries

Apparent Places of Fundamental Stars is prepared by the Astronomisches Rechen-Institut, Heidelberg (www.ari.uni-heidelberg.de). The printed version of APFS gives the data for a few fundamental stars only, together with the explanation and examples. The apparent places of stars using the FK6 or Hipparcos catalogues are provided by the on-line database ARIAPFS (www.ari.uni-heidelberg.de/ariapfs). The printed booklet also contains the so-called '10-Day-Stars' and the 'Circumpolar Stars' and is available from Verlag G. Braun, Karl-Friedrich-Strasse, 14–18, Karlsruhe, Germany.

Ephemerides of Minor Planets is prepared annually by the Institute of Applied Astronomy (www.ipa.nw.ru), and published by the Russian Academy of Sciences. Included in this volume are elements, opposition dates and opposition ephemerides of all numbered minor planets. This volume is available from the Institute of Theoretical Astronomy, Naberezhnaya Kutuzova 10, 191187 St. Petersburg, Russia.

Useful information and World Wide Web addresses on the Internet

Please refer to the relevant World Wide Web address for further details about the publications and services provided by the following organisations.

- H.M. Nautical Almanac Office at http://www.nao.rl.ac.uk/ & http://websurf.nao.rl.ac.uk/
- U.S. Naval Observatory at http://aa.usno.navy.mil/
- *The Astronomical Almanac Online* at http://asa.usno.navy.mil/ & http://asa.nao.rl.ac.uk/
- The Stationery Office (UK) at http://www.tso.co.uk/bookshop/
- U.S. Government Printing Office at http://www.access.gpo.gov/
- University Science Books at http://www.uscibooks.com/
- Willmann-Bell, PO Box 35025, Richmond VA 23235, USA, at http://www.willbell.com/
- Earth and Sky at http://www.earthandsky.co.uk/
- Macmillan Distribution at http://www.palgrave.com/
- Bernan Associates (TSO's agents in the U.S.) at http://www.bernan.com/